The Community and the Algorithm

A Digital Interactive Poetics

Edited by
Andrew Klobucar
New Jersey Institute of Technology

Series in Creative Writing Studies

VERNON PRESS

www.vernonpress.com

In the Americas:
Vernon Press
1000 N West Street, Suite 1200
Wilmington, Delaware, 19801
United States

In the rest of the world:
Vernon Press
C/Sancti Espiritu 17,
Malaga, 29006
Spain

Series in Creative Writing Studies

Library of Congress Control Number: 2021936400

ISBN: 978-1-62273-959-2

Cover design by Vernon Press. Cover image by Garik Barseghyan from Pixabay.

Table of contents

List of Figures and Tables

List of Figures

List of Tables

Foreword

What is this book about? Ostensibly, *computation and collaborative writing*. Esoterically? Collective embodiment. What is at stake here? Digital learning. Poetics. Possibly, the biosphere and humanity. What is key? Collaborating in untidy unity, with algorithms of cooperation, as the only path open, if sapiens is to survive and the Anthropocene is to be evaded.

Algorithms as communities. Society as play. Poetry as medicine. Learning as sharing. Inclusive creative education: *hacks, tips, tools, suggestions, and possibilities for optimizing* ***literary*** *life*. Community: 'care, accountability, and stewardship'.

Succinctly? Pedagogical Fluxus networked collaborative coded living systems communicating telepathic transparent creative, compassionate trust. That potential is the unanswered hope of techno-utopianism. Hybrid honey, hive mind.

Begin (as Andrew Klobucar - who has shepherded this compilation - states) *building an effective set of critical tools to help us engage, create, and dispute within a cybernetic state of unprecedented singularity.* Reimagine the literary classroom as a community inclusive of corporal and computational entities. Consider *swarm pedagogy*, bots of *humanist resistance*; *netprov* (inter**net** im**prov**isations) *binge play*; trans-mediated *text-sonification*; intimate *readings* of digital poetics; pathways *from satire to software*; *three-dimensional textual objects* in voice-activated VR; and a move *beyond zoom* into open source *collaborative thinking-writing*.

Each of the projects outlined/analyzed in this book originated from a collective, a cluster, a community; each was built using tools/code created by immense corporations and/or crowds. Every single cultural artefact, scientific theory or technological achievement is a *composite*. Just as our bodies are teeming modular repositories, so too are our digital tools. So too is language, a symbolic estuary of innovative impulses and ancient pressures; it carries us from isolated subjectivity into oceanic awareness; teaching emerges from its continuum.

Set aside career competition, it's time for new paradigms: cooperatism, collaboratism. Composites require collaborative sharing. Poetic permeability. Creative pedagogy is a crucial component in gestating a global civilization capable of recognizing earth (all its peoples, organisms, ecosystems and technologies) as a single cognitive-cardiac community circuit.

What if education were recognized as a public utility, a legally protected epistemological ecosystem implemented as open-source open-access non-profit github-wikipedia-etherpads? A version server of knowledge freely available, thriving with joyful contributions? Then perhaps society might avoid replicating old oligarchic hierarchical patterns of status and evolve altruistic shared repositories of luminous evolving helpful open-data.

Implicit within the notion of a *Community Algorithm* is a tension between crowds, computation, caring and critique. Critique originated as an unwillingness to be complicit (with corporations, branding, regimes, ideologies, hegemonies, etc.). Caring constitutes an empathic response, a shared vulnerability. Computation is communicative and analytic sinew. Connecting these threads, inviting crowds of readers, constitutes a relevant attempt to ensure that literature and literary education remains relevant and nourishing.

This epithet was written in spring 2021, as trillions of brood X cicadas emerged (distinct from the enigmatic internet *Cicada 3301: an edge of the field where the cicadas are silent, latent, timing us*). May their fascinating re-emergence remind us of cycles within time, inexorable explorations, and the power of instinct to re-ascend. Similarly, may the generous resonant ideas of community and cooperation contained explicitly and *subtly* in this book re-emerge to flourish, breed, and be of benefit. And finally, as is necessary within all relational ecosystems entering unknown territory, may we, as poetic-computation practitioners, recognize the necessity and power of forgiveness as an emotional subroutine enabling coherent communities to thrive.

David Jhave Johnston
Montreal, 05 2021

Acknowledgements

At a fundamental level, this book owes much of its focus to the growing range of research now looking at the influence of algorithmic programming on nearly all aspects of culture, whether we're producing or consuming it. References to current scholarship in what is popularly known as "algorithmic culture" appear throughout this very anthology in multiple papers, including my own; however, even more importantly, as a topic of investigation, the use of increasingly sophisticated algorithms to help shape and structure online cultural discourse continues to inspire live debates and workshop activities across the Humanities. I have been fortunate to participate in a variety of different forums, critical reading groups, as well as innovative platform launches, where artists and programmers work with each other to redefine art, performance, and writing in the digital era. As I note in my introductory essay, the seeds of this very project were sown at a media workshop I organized and directed with my wife, a learning designer for IBM, at the 2019 NeMLA convention held that spring in Washington D.C. At the time of completing this volume, that event still marks the last time NeMLA would be held at a single, physical location. There, I introduced my own writing platform The GTR Language Workbench, an online, drag and drop tool for generative text experiments alongside two other exciting media projects. Open workshops, as I had long come to learn, often prove to be the best format for bringing people together from a variety of different disciplines to explore exactly the kind of technology and practice-oriented issues featured in this anthology. Most of the chapters at some level, in fact, developed out of very similar interactive events, many of them designed as new and original technology demo sessions featuring highly innovative methods of use. I count among some of the more significant workshops that served as direct inspirations for this collection the always vibrant reading and writing series hosted month after month by The Kootenay School of Writing in Vancouver, Canada, as well as the vital Summer Institute offered every year by The Banff Centre for Arts and Creativity also in Canada. The Banff Centre has proven to be a unique and trusted resource for bringing researchers in the sciences and artists and critics from all disciplines together to work on independent projects, while engaging in stimulating dialogues and presentations. I attended one particularly important institute in the summer of 2005 with Workbench co-developer, David Ayre, developing the first programmable format of the tool as downloadable software. These events and the communities sponsoring them, compare well, I would also offer, to the thought-provoking media art and literary exhibitions held every year by The

Electronic Literature Organization as part of its annual conference. In general, showcasing media projects, whether in a single demo session before an audience or a more interactive workshop provides the dominant paradigm for most, if not all, discussions presented within the collection. The ongoing COVID-19 pandemic, and the havoc and fear it wrought throughout academic and creative communities over the past two years only made interactivity across digital networks seem more valuable as both a resource and format for engaging with one another and pursuing the dialogues we did. To this end, I want to recognize every contributor's individual effort to maintain an unparalleled intellectual community during this time, either together here, or with other equally respected organizations, sharing ideas, concerns, and encouragement. In numbers, there is strength!

I am especially indebted to the critical, often creative feedback I routinely received from my colleagues at the New Jersey Institute of Technology, including the Digital Humanities group, as led by its current executive team, Rosanna Dent, Gabrielle Esperdy, and Burcak Ozludil. For the past two years, they've organized NJIT's annual Digital Humanities Showcase, delivering another extraordinary arena for debate and discussion around programming, design and literature. During this same time, as I planned the anthology, I continued to examine several of its principal issues with friends and colleagues, many of them working in very similar areas of research and tool development, such as Louis Wells, a great friend and an accomplished innovator in the art of improvisation as both a genre of performance and method of instruction. Louis worked with me at the original 2019 NeMLA convention alongside his partner, Maria Aladren, who also contributed an important essay to this very collection. Other equally influential colleagues include poet, musician and long respected electronic literature critic, Chris T. Funkhouser, environmental ethicist and philosopher of technology, Eric Katz, writing analytics specialist Norbert Elliot, and Information Systems specialist Osama Eljabiri, who continues to organize multiple capstone opportunities every academic year for senior NJIT Computer Science students to engage with private and public technology initiatives on all levels, ranging from large corporate enterprises to smaller, independent experiments like my own. Before the pandemic hit our institutions of higher education along with the rest of the world, I was personally able to take advantage of this project to help further develop the Language Workbench's interface. All told, NJIT university, and in particular, its Humanities Department, has continued to support my projects and many similar ventures in digital media development as part of its core mandate. I am specifically indebted to the Department's organization of its Fall colloquium series in 2020 for enabling me to discuss more formally other recently developed writing interfaces and online tools like the Language Workbench now shaping our writing practices. I am thankful for the efforts of colleagues like Calista McRae and Vanessa Velez who showed much fortitude and a persistent generosity in maintaining the series through these extraordinary times. As with each set of seminars scheduled during an academic term, the Fall 2020 series gave me an opportunity to build an ultimately better, more collaborative writing workshop, this time using the now overly familiar technologies of video conferencing and remote learning that continue to dominate Higher Education across the country.

One indisputable fact available to some degree in every featured chapter is the enduring process of transformation literary studies and writing in general appear to be undergoing as fields of learning. Certainly, all writers in this collection would agree that the profound critical influence distributed networking and social media platforms seem to be having on contemporary writing practices did not begin with the pandemic. Such impact could be felt long before the SARS-CoV-2 virus had ever reached its first human host. Trends in technology that have been in development for decades are now merely accelerating to speed – perhaps even "warp speed." No doubt, serious research into what these trends portend for the literary arts, whether on or off-screen, is still in its infancy; yet, as I believe these essays show, some of the best critical insights have already been gained both spontaneously and collaboratively in the form of dynamic, often impromptu sandboxes, workshops, and interactive events. The past year has provided an unprecedented opportunity to actively engage and debate with colleagues around the world, screen to screen on questions we have had on our minds for at least two decades, often extending them into a variety of other fields, including programming, design, and visual presentation. In the year that has passed since the beginning of this project, we find ourselves in a startingly different media ecosystem. It is my sincere hope that these essays and the questions they raise provide useful starting points, perhaps a few loci of exploration, as we awake and step forward cautiously, yet bravely into our next new world.

Introduction.
The Community and the Algorithm: A Digital Interactive Poetics

Andrew Klobucar

New Jersey Institute of Technology

Abstract: The introductory chapter begins with a theoretical overview of how collaborative, online tools are shaping culture across a variety of disciplines and social practices. Collaboration has a rich history in 20th-century art and writing, often driven by innovations in communication and media technology that allow for better interactivity across distributed networks. Very recent advances in programming and algorithmic processing in digital media have led to an array of fascinating interactive experiments in both the arts and literature. The fact that many of the chapters were written at the onset of the COVID-19 pandemic only seems to ensure the cultural relevance of the observations and critical methodologies they offer. Taken together, the chapters show how computational thinking combined with faster, more sophisticated algorithms can generate a host of unique, interactive devices for collaborative, creative work in the digital world. The introduction's additional aim is to provide readers with a detailed summary of the unique array of techniques and interface designs featured in each of the chapters. Linking these works directly with many well-known historical projects in combinatory, generative writing and improvisational methods in the arts, this volume hopes to provide numerous points of open dialogue and discussion in order to help writers, readers, and educators build new communities of shared practices in a world where the rules of personal interaction have become increasingly unstable and seemingly subject to constant reinvention.

Keywords: Algorithms, Computational Thinking, Digital Media, Collaboration, Interactivity

***

Introduction

The Surrealist automatic writing exercise "the exquisite corpse shall drink the new wine," often abbreviated to "Exquisite Corpse" bears some renown among the many Surrealist collaborative art experiments. Its exact historical origins remain obscure, but the amusing diversion maintains an ongoing popularity as an enjoyable party game or collaborative classroom writing exercise that's easy to set up and quick to produce notable results regardless of how many contributors are involved. Its very title seems to testify to its ongoing mystique as a dependable method for generating captivating literary gems seemingly out of thin air. As one of the device's originators, André Breton, routinely noted in various early avant-garde writings, collaborative activities like games could be compared to dreams as modes for unlocking creativity and alternative psychic states of awareness. To hear Breton tell it, this particular Surrealist phrase appeared one night at a gathering of friends in Montparnasse in 1925, where he along with the artists Marcel Duchamps, Jacques Prévert, Yves Tanguy and Benjamin Péret adapted the rules of the parlor storytelling game "Consequences" and began to write, while concealing, random phrases on a sheet of folded paper. The first results were shockingly different than any single expectation, leading them to believe that an additional voice had somehow entered the room through their own conjuration or sorcery.

There remain today many variations derived from multiple game rules. In one version, each collaborator adds to a composition in sequence, either by following a generic syntax template, such as "article adjective noun adverb verb preposition article adjective noun," producing guided impromptu statements like "The grey book slowly burned beneath the glowing coals." Other versions allow each subsequent contributor to work directly from the last two or three words provided by the previous writer. Regardless of the different rules of play, a standard concept prevails. The game supports easy participation with the barest of tools: a single sheet of paper and enough pens or pencils for each player to use. After a set time, the contest almost inevitably results in a single, collaborative text that is usually intriguing to read by all contributors. Just as easily, players can elect to produce an illustration of some sort, where participants collectively attempt to draw a single figure while being constrained to produce only one portion of the work at a time. The first Surrealists, in fact, adapted the activity to include sketches soon after "Exquisite Corpse" became a standard exercise at collective meetings. Whether the work is a text or drawing, the procedural rules behind it follow the same pattern. Individual expression in an artwork is to be subjected to a set technique with the aim of encouraging a collaboration of sorts over the interests of any one person. No participant is able to see or read previous submissions when adding their own. The full manuscript or canvas is subsequently kept hidden from the group, revealing in the end a fantastic, usually fragmented set of lines or doodles to be read or seen together all at once as a coherent work of art.

The general appeal of this kind of activity doesn't seem complicated. Almost immediately, a messy, highly interactive group project takes shape, briefly shocking the contributors with its results, yet allowing each of them to escape full responsibility. Revelation inspires a particular delight in those who feel chosen and thus worthy enough to receive the final disclosure, especially having worked directly under the constraints of concealment. The works produced rarely compare with the literary expertise of a single writer working on a single set of ideas. Yet, a deeper significance seems to have been placed in the method, not the result. The work, in other words, is not so a message preconceived and then written, so much as a transmission duly received. A communication may have occurred, but the actual authoring now occurs on the receiving end, where the work appears fully encoded, ready to be deciphered. Breton, in fact, often compared the results to modes of fortune-telling, underscoring, perhaps, the movement's persistent interest in identifying some kind of metaphysical interaction with concrete reality as a core tenet. To explore what Breton termed the "Surreal" was not to abandon the concrete real for utter chaos, but to move beyond its supposed limits using alternative modes of reasoning. In his 1924 manifesto, he writes, "[i]t is not the fear of madness which will oblige us to leave the flag of imagination furled. … Surrealism is based on the belief in the superior reality of certain forms of previously neglected associations, in the omnipotence of dream, in the disinterested play of thought" (1969). Ludic or playful word association and the rules used to direct these language games quickly became a trusted mechanism for circumventing the standard, customary grammars that prevented the imagination from achieving its full potential as a source of creativity and "disinterested play." New constraints often designed ad hoc and for single occasions, Breton suggested, could lead to a "superior" realm of reason and whatever knowledge it might contain. This metaphysical rationale usually supported a collaborative, interactive approach to art and writing, allowing the Surrealists to exchange with each other strangely authoritative messages outside of any one participant's agency, immersing the final work in an aura of mystery.

Clearly, for Breton, part of the mystery derived from the prior feat of organising and imposing constraints on the individual tasks to follow, releasing each one from any preceding determination. Just how or why multiple moments of nonsense can lead to a superior, perhaps a strikingly imaginative layer of meaning always seemed to escape any subsequent attempt at fruitful analysis. Rather what amazed Breton was the broader perspective he could draw from the collection after its formulation. Thus lay "the omnipotence of dream," where captured fragments are set in motion to be executed as a new collective operation. The fact that such writing practices signified, not just a set of chance operations, but possibly their ultimate fulfillment underwrote the very idea of a grander Surrealist singularity. The specific constraints or rules one used to achieve this singularity should hardly matter, Breton reasoned; what was significant in these games was the fact that they could be executed, i.e., set in motion to generate something beyond or "over" our present, limited, individual sense of reality.

Breton's christening of the first such event an "Exquisite Corpse" was also more than a chance procedure. If we compare the execution of the program's main event to the pattern of countless prior myths when a rebirth of sorts is thought to loom on the horizon, a death had, in fact, been scheduled to take place that very first evening. For Breton, the renewal in question concerned modern art's ongoing evolution as a new set of aesthetic ideas and theories across multiple media formats. The death of realism, of course, had already been summoned in modernist painting a few decades earlier with the emergence of impressionism; hence surrealism was in some ways simply carrying that project to its next logical limit. The corpse of realism would now be entreated to its additional preparation and given a new wine – the power of an open-ended, generative process to allow modernism to achieve superior aesthetic ends.

It's not difficult to compare Breton's intense, almost spiritual relationship to language games and their play to the respectful awe many writers currently feel toward algorithmic procedures and their increasing prominence in everyday media platforms. In fact, there is growing political and cultural interest among media theorists, along with many computer programmers, in algorithmic processing in relation to knowledge construction in general. Rather than the epitome of what should seem to be an almost mathematically precise neutrality, the algorithm easily inspires an almost mystical sense of power and intimacy in coded language, allowing a single online program to scan far too deeply into how our own thought processes seem to be working at any one time. The results can be unsettling, as if, as media users, we have somehow conjured up something much more malevolent than a lost Ouija spirit lurking amidst our Java syntax.

Media networks, whether based in the world of art or commerce, including high-tech start-ups, continue to foster collaborative modes of engagement, often highlighting the value of interdependent work environments over more traditional hierarchical arrangements. Just as significant, however, is the capacity of distributed networks to sponsor and better manage such activities through programmable, data-driven algorithms and computation in general.

Once employed, as demonstrated as early as Breton's original Surrealist meetings, algorithms may usefully machinate individual activities into a more unified, collective format without automatically denying us the promise of our separate imaginations, possibly even a dream or two. All that seems certain once networks become more of a standard structure for creative media practice is that the various ways or means we engage our imagination, whether collectively or individually, does seem significantly transformed. Throughout this volume, the precise methods and models being employed still evoke a certain level of mystery regarding how they may actually play out, despite how intricately detailed any algorithms in use happen to be. In many cases, the persistent sense of enigma haunting many collaborative, computational media experiments in terms of how they operate might even be enhanced by this intricacy. The innovation and creativity these networked projects sponsor seem plainly evident. One important question this volume seeks to address is whether such networks can actually help establish new forms of communities or communal environments outside individual cultural events, and if so, how might these communities function in an increasingly computation-driven, media-enhanced society.

The rapid postwar development of distributed network technologies and its influence on emerging counterculture art movements in the 1960s, especially in new media, remain a popular area of focus in art history and cultural criticism. Together, these two advances in media ecosystems provide a useful, perhaps even essential, context for many of the current advances shaping digital platforms in art and writing. Of specific note here are groups like the art media collective USCO (US Company) and Robert Filliou's and George Brecht's unique Fluxus-based experiment "The Eternal Network," both of which sought to bring together important social and even participatory elements into new media projects. In tandem with these art media projects, throughout the 1960s, similar experiments in literature, inspired, in part, by Breton and Surrealism, attempted to merge computation more formally with writing, as we see with poets like Brion Gysin, Jackson Mac Low, and, of course, members of the French artists collective "Oeuvres Litteraire Potential" or OULIPO. The "internet" in concept was still little more than a thought experiment to improve ways to share computer resources initiated by the United States Department of Defense, yet what is historically significant in the early 1960s is the growing cultural interest among artists and philosophers to reimagine computation as a mode of networked, creative collaboration. Studies of the historical and cultural dimensions of algorithmic processing cannot ignore the significance of distributed network technology and its development over the last 60 years. Once computer systems could be assembled into multi-terminal networks with more computational power as was possible at the beginning of the 1960s, it seems that nothing less than a new paradigm of electronic communications emerged.

My own work and interest in developing more socially interactive digital writing tools for live collaborations in text generation and combinatory poetics owes much inspiration to each of these earlier projects and others like them. USCO and The Eternal Network demonstrate together a broader critical interest in shifting part of the focus in postwar creative media experiments from the art object itself to the actual process of its production, including the various roles artists and writers may take on when conceiving and developing a specific work for display. Filliou's and Brecht's vision of contemporary art practices in terms of an "eternal network" of shared ideas and resources, supplemented with effective feedback, did more than outline a more economically advanced distribution model; it actually proposed a fundamental re-conception of creativity as an interactive effort. Communication, these "networkers" eternally theorized, the capacity to connect people, not individual skill, would be the true source of imaginative invention regardless of genre or type of art materials in use. The primary communication model the two artists typically referred to when qualifying the sheer power they felt collaboration and interactivity in art could fuel was, in fact, telepathy.

As fantastical as telepathic intuition may seem as a mode of communication, the process speaks plainly to the level of intimacy and interpersonal trust Filliou and Brecht considered inherent in all creative practices. For Filliou specifically, as long as art emphasized individual effort over more collaborative gestures, it would remain tied to competitive struggles for personal primacy and cultural dominance. Artists routinely sabotaged each other as part of the profession, securing their aims to the rules of markets rather than the good of the project. Telepathy, he thought, offered a useful critique of this cultural model where reading minds became a kind of metaphor for a truly transparent social economy based upon honesty and cooperation rather than aggressive claims to authority.

André Breton did not likely consider Surrealist collaborations formal exercises in networking or even building a new kind of social collective. There is no record of him using such terms in any of his theoretical works; as a writing procedure, automatism was understood more in terms of psychological state, not a social one. Yet, clearly, by the time media projects like the Eternal Network and USCO are in play, we can see a much broader cultural interest to explore networking as a formal model for transforming individual art practices into collective structures. Computer programming, coding and networking developed into highly creative cultural practices in their own right. As well, the prioritization of technique and procedural constraints within these working collectives over individual expression aligns such initiatives with many core features of modernism in general. From both a technological and modernist perspective, networks might best be understood as a formal means of communication independent of any message content. Key to this relationship, in fact, is the unique, often prodigious social tension it can generate between the combined efforts of collectives and those of individuals. In other words, the network showcases collective structures almost paradigmatically as living systems or constantly evolving environments. Networked collaboration, as we see throughout modernist formalism, is built upon an active in-process struggle to find structural harmony, and not simply a demand for personal surrender to whatever arrangement appears to be in play. The fact that such harmony remains typically elusive in the modernist form is actually one of the movement's core aesthetic strengths.

Computer programming evokes many of these same issues, especially in terms of designing collaborative models of interaction as living processes of social engagement. Chapter 8 of this anthology details the important alignments in programming, networking and modernist aesthetics that guided my own joint experiments with the algorithm as a collaborative writing tool over the past decade. The platform I co-developed with programmer and poet David Ayre, "The GTR Language Workbench," provides an online drag and drop interface to perform numerous algorithmic processes on texts, including several surrealist language games, allowing writers to explore and learn about combinatoric literature on screen. As with many of the contributors to this volume, programming gave me the opportunity as a teacher and literary critic working almost exclusively with digital media to investigate how algorithmic operations might be used to "network" creatively and credibly across many different writing formats. Our Workbench platform with its capacity to generate and make available an array of processors for collaborative writing exercises draws historically from the media experiments mentioned above, including Breton's own exquisite resurrection of the text via new wine. Other processors draw directly from various OULIPO language games, such as Jean Lescure's N+7 method where parts of speech in a source text, such as nouns, verbs, or adjectives, are replaced with counterparts taken from a standard dictionary according to a preset numbered position – usually seven words up or down the list. The resulting text will usually retain the general tone and meaning of the original, especially if the position variable is kept low, with multiple eccentric additions. The technical sophistication of this type of algorithmic instruction seems hardly comparable to the digital semantic technologies that have become available this past decade. Advances in online lexical databases like Wordnet and other text analysis tools, as well as Natural Language Processing (NLP) in general, have drastically transformed online writing. Scholarship in the relatively new academic field of the Digital Humanities continues to develop its own methodologies and resources in direct response to the expanding significance of computation throughout the liberal arts. At the same time, the reconceptualization of the network as a creative, collaborative exercise in media art demonstrates how central algorithmic design has been to cultural modernism since its beginnings.

Several writing projects featured here, namely, Maria Aladren's Swarm techniques for teaching and Maria Lantin's experiments with VR and sound were presented alongside my own online platform, The GTR Language Workbench, in a special workshop hosted by the 2019 Northeast annual regional MLA conference (NeMLA), focused specifically on questions of networking, collaboration, and the algorithm in teaching writing. I organized the workshop together with Sharla Sava, the author of Chapter 8's more complete history of "The GTR Workbench," to explore various approaches to designing in-class collaborative exercises in online media as well as provide more research into how networks and networking have influenced developments in media art through the postwar era. Each presentation in the NeMLA workshop consistently demonstrated that software code and its execution rarely exemplified its stereotypical association with calculated precision and rigorous design; instead, the art of programming seems better understood as an ongoing, perhaps even interminable process of trial and error-driven by constant collaborative feedback.

Digital Utopianism and the *Whole Earth Catalog*

Despite the creative potential evident in early postwar forays into media art and programming, further research into distributed network culture also indicates that many of these same initiatives to foster collaboration and process-based interaction in art practices often succumbed to a much more problematic techno-utopianism by the late 1960s. Most studies of this historical tendency tend to showcase the WELL/*Whole Earth Catalog* movement as its clearest archetype. The *Whole Earth Catalog*, produced initially as a kind of contemporary almanac or newsletter for independently minded artists and thinkers, quickly became an international symbol for a new kind of technology-based cultural mindset. However, even when Stewart Brand began his publication in 1968, there is little doubt that he imagined he was setting in place nothing less than a community cornerstone with its own ideology. Brand, himself, declared in the inaugural issue that the periodical would foster communication and the sharing of "hacks, tips, tools, suggestions, and possibilities for optimizing your life" (Colman, 2011). While the projects explored throughout this collection continue to underscore computation and networking as process-based techniques for collaborative thinking, it seems suitable here to review how an unquestioned cultural conviction in digital technology's capacity to deliver rationalist, data-led worldviews has also developed from very similar collective initiatives in media art and writing. Historians agree that it remains no exaggeration to characterize *Whole Earth Catalog*'s general readership as a very successful and unified cultural movement due to the shared sensibility it explicitly promoted. To subscribe to this tool, as Brand called it, was to openly profess a profound dissatisfaction in mainstream culture, whether delivered via government, the education system, or the private sector. Instead, the *Whole Earth Catalog* promised access to superior, alternative sources of information and knowledge. Rui Torres and Daniela Côrtes Maduro in their chapter "Wreadings: Digital Poetry and Collaborative Practice" draw an important historical line between the original *Catalog* and the first complete hypertext work of fiction, Judy Malloy's network-as-novel *Uncle Roger* (1986-), crediting its original design and publication to the Electronic BBS Art Com Electronic Network, a project associated with the *Catalog*'s later digital version, the WELL. Yet, as progressive as this pronouncement may have seemed at the time, studies like Janet Kraynak's recent book, *Contemporary Art and the Digitization of Everyday Life* (2020), argue that Brand's *Catalog* exemplifies how countercultural perspectives on media networking and collaboration abandoned interests in political contention and live debate for an array of utopian precepts and ideological positivisms (Kraynak, 2020). For her, projects like the *Whole Earth Catalogue*, although progressive in its attempt to apply information and media tools to art practices, also sponsored a much broader compromise between core democratic precepts and the neoliberal, technocratic model of social networking that flourished between the 1970s and the first decade of the current century. Fred Turner's important 2006 critical history of the *Whole Earth Catalog, From Counterculture to Cyberculture*, appearing just before the 2008 financial crash, provides a similar overview of how media networks fostered ideologies that, in turn, contributed to current economic and social developments in digital media that appear simultaneously capable of promoting and inhibiting progressive calls for action in the name of democracy. During the time of social protests in the mid-1960s, almost 30 years before the emergence of the World Wide Web, he reminds his readers, "computers had been the tools and emblems of the same unfeeling industrial-era social machine whose collapse they now seemed ready to bring about" (1). This persistent duality in both form and function remains a fundamental point of contention in contemporary considerations of computational culture and digital media. Social perspectives across a variety of contexts easily deem algorithms as almost universal symbols of objectivity and precision; at the same time, their growing use in the arts and social sciences demonstrates highly complex, often unpredictable processes constantly subject to the demands of collective input and whatever instances of cultural bias, data error, and, at times, simply blind faith that may accompany this input.

Our workshop on computation and collaborative writing in the digital classroom carried forward similar questions and concerns regarding programming as a cultural practice, drawing from many prior experiments in media art and the examples of creative play and interactive dialogue these projects generated. It is in the algorithm's capacity to fuel active debate in the pursuit of coherence, we argued, that gave the algorithm its cultural significance, not its mythological status as some imagined pinnacle of empirical objectivity. Not surprisingly, as algorithmic computation continues to permeate more and more modes of cultural interaction, an increasing number of studies also makes clear its many inherent flaws as a source of unquestioned certainty. The recent work of researchers like Safiya Umoja Noble (*Algorithms of Oppression*, 2018) and Cathy O'Neil (*Weapons of Math Destruction*, 2016) show that data routinely used for machine learning inevitably reflect existing social and political predispositions in many subsequent outcomes and cannot for this reason be considered pure results of applied formula. Companies like Predictive Policing (PredPol®) currently use algorithms to recommend where best to allocate police and security forces in neighbourhoods anticipated as being vulnerable to criminal activity. Their target market consists primarily of companies in commercial districts ready to subscribe to information services that show with supposed algorithmic precision where potential liabilities may exist. PredPol®'s technology page shows the algorithm it uses, listing the behaviour variables inputted into different equations: namely, instances of "repeat victimization", "near-repeat victimization, and "local search" data (PredPol®). It's not surprising to suppose that private security companies will become increasingly dependent on this type of market. If, however, as both Noble and O'Neil maintain, the reporting and inputting of criminal activity is racially biased, the corresponding results PredPol® algorithms determine cannot possibly be as objective as they claim. At the very least, critics also point out, predictions listing specific geographic locations will likely discourage initiatives in planned financial and social investment needed to improve these same locations over time, thus creating the very conditions the algorithms originally predicted. Essays in the first "Foundations" section of this anthology, such as Charles Baldwin's intriguing narrative "Cicada" and Taeyoon Choi's thoughtful "Community Code: A thing, something, everything and nothing" detail some of the more critical cultural effects directly related to the growing influence of algorithmic computation on social interaction, many of them rooted in an unquestioned faith in information technology as a means to objective knowledge.

Computation and Late Finance Capital

An expanding range of critical research over the past five years continues to argue for a much more complex, often less secure model of how algorithms work. And yet, the more problematic stereotype of computational reason as a device of objective precision still persists. In fact, the very foundation of late finance capital in many ways continues to herald the power of the algorithm as a progressive, ultimately more exact as well as exacting device for major financial decisions. Algorithmic models of consumer activity, some of them openly proposing alternatives to traditional industrial models of programmatic utility in social interaction, continue to be popular across many new markets. Alex Rosenblat's recent work on the algorithm and labour, *Uberland: How Algorithms are Rewriting the Rules of Work* (2018), contends that some of today's most successful technology companies have "created a fundamental shift in what it means to be employed" across the globe (15). Rosenblat does not attribute this development to any specific advances in hardware, despite the innovations we see in robotics and media tools now enhancing industrial production. The most significant change, as the title makes clear, derives directly from algorithmic computation due to its unparalleled capacity to redefine market relations from both the consumer's and producer's perspective. The prototypical company for Rosenblat's analysis is, of course, Uber, a transportation company that has somehow managed to redefine itself as a completely new kind of consumer service with almost no employees. This unique transformation, it seems, is due primarily to Uber's creative use of algorithmic technology as the essential tool or device being purchased by both drivers and their customers from an equal position to use as they see fit. Whereas a customer in need of a driver may subscribe to the service to find the ride they need at an apt time, the driver also subscribes to the same amenity to find the ride they desire to serve with equal utility. The profound economic shift Rosenblat refers to for his core thesis derives, it seems, from the algorithm's unprecedented economic capacity to eliminate employees, not by downsizing, but rather by making everyone a consumer.

As Rosenblat makes clear, algorithmic processes do more than simply increase market growth; they have initiated changes in the very structure of the modern social economy. Using computation to rationalize consumption patterns, late finance capitalism presents perhaps less prospects for productive labour in this economy, but clearly more opportunities to preview our purchasing habits and respond to them with greater and greater accuracy. An algorithmic market grows by surveilling and then managing increasingly personal – even intimate – aspects of our private and professional lives. The rapid spread of these technologies into nearly all aspects of our culture prompts other critics like Jonathan Roberge and Robert Seyfert to consider algorithmic thinking itself as one of the primary discourses guiding 21st century social relations. On one level, it has allowed a more flexible, service-based consumer economy to emerge across traditional class and race identities. At the same time, the intrusion of the algorithm into capital markets has led to the gradual erosion of these very same identities which in turn brings on feelings of community disorientation and personal isolation. In their recent anthology of essays *Algorithmic cultures: Essays on Meaning, Performance, and New Technologies* (2016), Roberge and Seyfert begin by noting that "[t]he current, widespread dissemination of algorithms represents a double challenge for both our society and the social sciences tasked with studying and making sense of them. Algorithms have expanded and woven their logic into the very fabric of all social processes, interactions and experiences that increasingly hinge on computation to unfold" (1). To understand algorithms both as a tool and an abstract concept, Roberge and Seyfert cite the logician and computer scientist Robert Kowalski's original description of them as mechanisms of "logic and control" (1979). A better grasp of their ideological and economic roles, however, requires one to look deeply beyond their initial relationship to sequential ordering and its regulation. When traditional social spaces of interaction and cultural identity, like different neighborhoods, public city zones, and even our home environments are being managed computationally – usually to maintain constant real estate growth and quick market turnovers – an entirely new algorithmic community begins to emerge. In this community, social interaction is not a natural outcome of organic neighborhood development but is instead the result of finely tuned economic constraints executed by an even larger programming apparatus. Whatever capabilities this apparatus may demonstrate for control, its faculties for automatically determining equity and justice as social aims seem limited.

Theorists like David Harvey have long provided important critiques of the postwar political economy's failure to provide proper government oversight for algorithmic markets, qualifying this breakdown specifically as a central ideological feature of neoliberalism. In Harvey's view, advances in information technology from the 1970s onward are unequivocally designed to "bring all human action into the domain of the market" driven primarily by our growing digital "capacities to accumulate, store, transfer, analyze, and use massive databases to guide decisions in the global marketplace" (3). We may tacitly accept our predetermined role in this new ecosystem as one of its growing number of consumers/employees; yet it can still seem especially challenging to gather a working perspective on just how these technologies may be transforming our own cultural relations, never mind the underlying economic structure driving the many transitions we see around us. For artists, teachers, and even coders, the tools we use, the increasingly complex media resources now indispensable to our reading and writing practices frequently appear wholly alien in terms of how they are shaping our interactions and general ability to communicate with each other. All that seems certain is that we are in the midst of an unprecedented, possibly interminable state of evolution. In fact, one of the most consistent observations appearing throughout this collection ardently acknowledges how little we still know as writers and readers about how algorithms function as part of digital communication networks.

From Surrealism onward through many modernist movements, the lack of predetermined aims within a collaborative, networked approach to media artworks remains a defining aesthetic feature. To adopt such procedures is to accept various liberating features that rules and constraints can bring to creative projects. The persistent opacity in how many algorithms often seem to function may also testify to their capacity to move beyond individual interests and personal observations in any artistic endeavour. The circuitous pathways of interactivity and exchange they exhibit also reveal countless procedural options for a variety of different data patterns and objectives. Certainly, Kowalski's emphasis on logic and control remains central to an algorithm's design, but, as we'll see throughout this volume, key to their overall complexity in form and structure is their capacity to collect and organize multiple points of input into meaningful transmissions. At the same time, algorithms can all too easily convince their users, whether they be artists or financial managers, into imagining new levels of universal meaning offered by objective calculation. Identify patterns and functions via algorithms, and suddenly any number of mysterious affiliations, both organic and inorganic, suddenly seem possible. As one critic, Ed Finn, points out in his study *What Algorithms Want* (2017), taken together, the constraints and rules that make up algorithms might best be thought of as a "critical frame that is interpreting objects which is also interpreting you" (55). Clearly, both the structure and utility of algorithms present together two fascinatingly distinct forms: on one hand, we have a novel technique for deciphering persistent quandaries by determining larger patterns and combinations; at the same time, we have powerful tools for rendering new objects, new "boxes" that, in turn, must also be similarly deciphered. Finn also usefully lays out three levels of algorithmic activity according to the intricacy of the rules and constraints in operation. If we assign algorithmic activity at its most basic level to any sequential procedure used to determine a single preset aim or goal, then multiple, more variable aims automatically infer more elaborate algorithm designs. True to form, as these designs become even more complex, the constraints and rules guiding them are able to execute an increasing number of unique outcomes above and beyond any original purposes. Increase the level of interactive intricacy in an algorithm's configuration, and suddenly its output begins to resemble intelligent, reasoned determinations.

With so many different levels of interaction possible, it's not surprising that the algorithm continues to confound any easy theorizations. Its devices seem to embody while simultaneously rebuffing pure precision in rational thought. Shintaro Miyazaki's recent essay "Algorhythmic Ecosystems: Neoliberal Couplings and Their Pathogenesis 1960-Present" (2016) seems to capture some of the duality we see in their simultaneous use as both a creative networking tool and a method of decipherment. The essay begins by recalling the algorithm's ongoing economic function in our current era of late finance capitalism, where social relations remain inexorably dependent upon complex, time-based interactions with built-in response mechanisms to ensure aim-driven efficiency and economic profit. An alternative ecosystem to neoliberalism, Miyazaki argues, reconsiders computation as a mode of physical "signal processing," downplaying traditional associations with calculation in favor of dynamic feedback and constant interaction. Put another way, the algorithm isn't confined to specific aims or goals in Miyazaki's view; rather, it is an ongoing process, materially generating new practices of learning and conceptualization. Working with algorithms, thus, almost inherently involves collaboration, especially given the importance of reflection and response when pursuing cognitive growth. As Miyazaki notes, to understand algorithmic computation primarily in terms of "signal processing", aligning it with the "rhythms of 'spacetimematterings'," is to move it beyond its more typical restrictions as a time-based sequence of ordered calculation (129). The term Miyazaki uses to qualify his sense of meaning as a transmitted pattern arrives by way of feminist theorist and quantum physicist Karen Barad. Barad conjured the phrase out of quantum physics in order to underscore precisely how a quantum-based understanding of reality might help us build a more progressive, if not accurate, materialism in terms of ontology and epistemology. For Barad, quantum theory's unique conception of objects, in fact, matter, itself, calls into being a reality that is by nature physically indefinite, demanding, in turn, a much more open, less categorically fixed sense of individual agency. Bereft of any last word on how reality actually functions, Barad argues, no individual or social group can place their perspective as scientifically superior over another. With this model, a revised social ethos, where whatever role humans play in controlling or even identifying causation in the universe, seems better understood as a part of a broader entanglement of multiple material forces. Miyazaki uses the term similarly to acknowledge the many indeterminable, impermanent aspects of algorithmic meaning. No single device prioritizes one conclusion over another: "[M]achines not only speak or watch one other, but rather – to formulate it in a more technologically accurate way – listen for and detect one another's signals and their rhythms" (129). The very patterns of computational processes, it seems, bear a certain level of meaning for the machines producing them distinct from any quantified information reserved for their human counterparts. Most significantly, these rhythms may also avoid the political implications discussed earlier since they occur outside preset algorithmic instructions and biases guiding the input of the original data. Miyazaki's creative algorhythmic model helps us reimagine computational interfaces as highly complex, socially collaborative modes of interaction, and not merely a means of systemic instruction to acquire information. Both formally and physically, the screen seems to transform into a much more dynamic, if labyrinthine, database of media files and synchronous interactivity. In effect, we have learned via these devices to read anew with our fingertips as modes of collaborative processing. In Miyazaki's words, "[w]hen an algorithm is executed, processes of transformation, and of transduction from the mathematical realm into physical reality, are involved. These processes are not trivial. They have been designed to appear simple, but the becoming of an algorithm, its unfolding and metamorphosis into an algorhythm, often involves issues, problems, frictions and breakdowns" (135). The concept of information as rhythm and flow becomes easier to imagine when it occurs digitally given how instantaneous its ongoing transformation can seem. The result of this rhythm seems also well described here as an unfolding, invoking the same style of gradual revelation enacted in Surrealist language games like "Exquisite Corpse". Unfoldings in many ways capture how algorithms tend to convey their results, few of them being singularly determined by fixed rules or routines. Algorithmic computations unfold gradually, as data transmit and transform through procedures, patterns emerging via internally generated rhythms.

Computation, Conflict, and Collaboration

Distributed networks and their continued development over the past half-century have resulted in a startlingly rich field of cultural analysis, ranging in part from the social history of Douglas Kellner to the explorative, turn-of-the-century work on hypertext media by writers like George P. Landow, Jay David Bolter, and Matthew Kirschenbaum, as well as the more experimental cybernetics of Sadie Plant and Donna Haraway. Current media theory, as we've seen, continues to focus on the rapid evolution of digital interface technologies and their ongoing transformation of telecommunication, while emphasizing the consistent advance of an algorithm-driven, diffusion-based relationship to culture in general. One common feature shared between theories of distributed networking and algorithmic processing remains the inherently fluid nature of any computational environment operating as an open system, which is to say as a system with feedback or response mechanisms. Algorithmic processing, consistent with distributed networks in general, is built around system routines able to manage data with a near-endless variability in form or content. The algorithm, like computation in general, may have emerged culturally as a tool associated with control and monological reasoning; but it quickly came to signify more organically multifarious modes of analysis.

Two related fields of study recently developed within the Digital Humanities, electronic literary criticism and software studies, continue to advance research into networks and network technology as cultural phenomena. As writing and reading practices, as well as distribution methods, continue to adopt interactive, network-based platforms for more diverse, digital audiences as a primary literary apparatus, it seems increasingly relevant to cultivate more creative and critical relationships between programmers and authors in general. Electronic literature (e-lit) critic Leonardo Flores considers social media networks distinct to an entire new generation of media artists and authors emerging in the second decade of the 21st century. Calling them a third wave of innovators, Flores decides the current online environment has transformed digital authorship beyond its initial pre-web stage leading up to the mid-1990s, as well as all web-based experiments appearing before 2010. For Flores, once platforms and network-based apps are adopted in literature, an unprecedented wide range of electronic literary works “such as bots, electronic poetry, videopoetry, hypertext fiction, mobile and locative works, virtual reality, augmented reality ... are revitalized as they find new forms” (2019). However, these new relationships between computation and the literary arts are ultimately determined, network technology clearly remains a key formative factor. Cultural theorist Christopher Vitale, in fact, looks specifically at networks as the basis for a new philosophy in his 2014 book *Networkologies* in order to develop a working set of principles and possible viewpoints based on their structure. Fittingly, distributed networking continues to supply one of the more effective paradigms for understanding collaborative, computationally driven literary projects and cultural theory. Flores's work in electronic literature is particularly useful for expanding the literary arts by focusing on individual works in interactive fiction, whether they are categorized in any one of the three generations he lays out. Looking at the third, most recent age, we are invited to apply many of the more traditional literary tools of criticism, including the practice of close reading, to devices like Twitter bots and poetry generators attached to the online neural net software. Software studies overall follow a similar approach, respecting programs and even single lines of code as independent works available for critical readings. At the same time, the field can benefit from the philosophical perspective scholars like Vitale provides in order to understand software and programming practices within broader epistemologies and ontological frameworks much as prior lineages of modern reasoning help situate analogue technologies. While acknowledging these methods, this collection seeks to diversify critical thinking in digital media even further by offering an additional set of interrelated methods of analysis for recent network-based, computational art and writing created in the wake of Web 2.0. The volume begins by outlining a set of theoretical foundations, before exploring recent pedagogical concerns and practices, and concludes with several case histories. As one would expect, no chapter in this volume can be exclusively contained in a single section. Each contribution is grounded in one or more theoretical discussions in digital media, while many have the potential to be used directly in the classroom. The breadth of topics covered by the chapters seem relevant to a variety of courses, ranging from studies in digital culture and humanistic computing to media art studios. At the same time, all contributions draw from specific case reports, both historical and contemporary, underscoring the significance of collaborative networking in art as a live and ongoing practice.

One further element directly connecting each of the sections together, aside from the actual technology required for network-based media art, is clearly the current pandemic crisis. Most of the individual contributions emphasize the general importance of collaborative effort in the process of building understanding and support across different communities during times of extraordinary social change. One might view the very introduction of computation into domestic living spaces between 1970 and 1990 as precisely one such phase. No doubt, the effect the pandemic continues to have on how we communicate and work with one another will find its place in history as a similarly important period of media reinvention. Anastasia Salter, Taeyoon Choi, and Anna Nacher, in fact, all identify COVID-19 as a specific reason why their ideas and the media technologies they reference are especially relevant to educators and artists at this point in time. Salter explicitly reminds us that many past trends in the use of digital media in classrooms will inevitably be re-examined and perhaps placed under more intense inspection given the threat COVID-19 now presents to traditional pedagogical practices; Choi summarizes the extensive social effects of the pandemic quite literally "as a chance to reset our Community Code," while Nacher declares networked writing spaces to be a new "staple of professional and private life" now that COVID-19 has driven much of the world's population into a semi-permanent state of domestic lock-down. Upon first reflection, the new media networks sponsoring the economic changes of late finance capitalism may hardly seem comparable to those preserving social engagement during a time of worldwide contagion. Yet, as these three authors exemplify, digital media artists and writers have considered algorithmic networking vital to defining a more inclusive poetics of communication and information exchange across our ecosystem for almost half a century. The networks we see in front of us were begun long before Uber drivers hit the streets armed with algorithmic contact information and artists and educators began to imagine social space primarily as a network of screens.

Taken together, the three sections organising this collection show a transformative moment in networked, personal computing, as our political economy continues to adapt to near-global, instantaneous smartphone interaction. The metamorphosis of the smartphone over the past decade into an increasingly intimate personal apparatus, often synchronized with wrist-based fitness devices like FitBit or the Apple watch, paves the way for the next phase of distributed digital networking as a form of biotechnology. Both Nacher and Choi seem to agree that once digital networking and information exchange begins to function as a biotechnology, the resulting ecosystem conjures a newly collective sense of agency that compares well to a social body under the pervasive threat of worldwide contagion. Nacher very clearly reminds us how easily any social threats this networked agency carries may be misidentified and accepted as just another promising resource: "What happens," she asks, "when an algorithmic entity – however ambiguous and entangled in the logics of corporate digital capitalism and intensive, profit- and metrics-oriented data tracing – becomes the part of a treasured relationship?" It's tempting to re-imagine this relationship as asymptomatic at an individual level, yet highly destructive to the community. At the same time, the sheer variety of perspectives available here introduce us to an impressive array of different technological approaches to understanding and working within this ecosystem. The tools discussed range from the relatively simple use of shared Google documents to the application of VR as a highly advanced mode of interactive composition. One might even compare this growing diversity in tactics and their varying critical capacities to analyze our media environments to the numerous inoculation methods the pandemic continues to fuel. Such an array demonstrates just how complex and diverse our current attempts to build a working poetics of communication and information exchange for this new era need to be. Rob Wittig's and Mark Marino's collaborative "NetProv" project, outlined by Wittig in Chapter 6, combines gaming with theatre using different social media platforms to enact live group performances. Following the political and economic success of popular media brands like Twitter and Facebook, NetProv has only grown in cultural relevance over the past decade with each subsequent rendition. Wittig's essay provides a clear historical overview of the NetProv project before introducing helpful methods for different groups in almost any context to generate similar media experiments of their own. One of the strengths of NetProv theatre is its adaptability; the real core ingredient in each project, however, is collaboration. To be creative in a NetProv project demands, foremost, a willingness to work with others. In some ways, NetProv seems to exemplify how a functioning social network can build creativity out of the straightforward act of conversing with others. Wittig and Marino plainly imagine NetProv as an art of social interaction, where algorithms become tools for assimilating, scripting, and giving shape to any number of spontaneous dialogues and discussions. Each NetProv event can be seen in itself as a type of algorithmic device, providing rules and controlled tasks for a specific technical platform to be used by players and performers alike. The performance also takes the form of an ongoing media transmission broadcasted over an extended period of time. In fact, once the featured platform in each project becomes a broadcast event, social media interaction seems to fall into place as an entirely new genre of collaborative performance art.

The aesthetics of digital media collaboration is also apparent in projects discussed by Maria Aladren ("Digital Swarm Techniques") and Maria Lantin and Simon L. Overstall's case study of their own impressive VR work, *I am Afraid.* In some ways, Lantin's and Overstall's project complements Wittig's experiments with Netprov, differing primarily with the level of programming and type of hardware they've developed. Exploring the possibilities of "Voice as Sonic Sculpture," the authors introduce numerous VR environments where participants interact in real time using different applications to design and perform a sound poetry event together. The title piece, *I Am Afraid,* specifically invites players to compose various "sonic objects" using their voices to speak into a virtual, yet communal space. The resulting objects immediately reassemble into a single co-authored text, as the software translates spoken words and phrases into floating images of sound. On one level, the technology shows us how easily high-speed network connections and advanced software can now transform individual media input across time and space into live events. In fact, as writing tools, the devices seem to exemplify what Walter J. Ong notably phrased the "technologizing of the word." In this environment, however, the resulting text seems to re-discover much of its earlier oral value as sound, while at the same time taking the paradigm even further as digital information coded as VR imagery. If the earlier transition from oral sound to print transformed language into a more exclusive, private, silent experience, the VR tool by mediating individual contributions as part of a communal environment subsequently seems to rediscover orality's inherent value as an explicitly public, audible event. Fear, as *I Am Afraid* just as candidly implies, is a key sonic, not to mention psychological, attribute in the work, inviting us to question the emotional health of technological environments as individually interior experiences. *I Am Afraid* performs a number of related psychological gestures, signifying emotional confusion, anxiety, and even sadness by sounding out what the authors call "fear-induced words." It's tempting to interpret this sonic event somewhat therapeutically, where composers are able to share and thus possibly externalize states of internal distress. Once again, signals caught in midst of being processed or transmitted figure prominently in these performances as a unique aesthetic event. Recalling Miyazaki's insight into computational interactivity, here, too, algorithms appear to guide the flow of data from one device to another in an imaginary, if temporary, media continuum.

A different kind of text-as-sound event is also demonstrated in the "Singling" project developed and presented by Kedrick James, Esteban Morales, Rachel Horst, and Effiam Yung as part of a new field of study they identify as "literacoustics". Comparable to Lantin and Overstall's project, Singling also acknowledges a kind of "second orality," where, with reference to Ong's theory, digital texts eschew private modes of silent reading for a more public performance of a text as a sonic or sound-based transmission. In some ways, even more explicitly than the VR event *I Am Afraid*, Singling's literacoustic transmissions demonstrate sonic signals as an active, live creation. The sounds being generated are not just audio representations of human voices, but instead, using MIDI protocol, are able to program instructions for different sonic patterns or rhythms in real-time. Algorithms operate here quite literally as the means to move one set of data from one form or format to another. The transformations subsequently in play often seem appropriately disquieting to experience.

While tools like Singling and the VR applications developed by Lantin and Overstall, show highly sophisticated programming practices, other devices like Maria Aladren's "Swarm techniques" underscore just how intrinsic networking as a methodology has become in our communication practices independent of any specific hardware. For Aladren, online networks like the web quickly become useful metaphors for any collaborative space of interactivity. Using methods drawn in part from non-hierarchical improvisation theory, Aladren reimagines social networking online as an effective, often overlooked resource for building shared creative environments at a moment's notice. The procedures she outlines easily recall the original Surrealist enthusiasm for collaborative writing described earlier. When writing occurs synchronously and collectively across the internet, the resulting texts seem best characterized, she argues, as an unending performance, shedding any allusion to planned projects with predetermined goals. The various exercises Aladren introduces often need little technical expertise beyond solid access to a tool like Google Docs. Instead, the primary emphasis within these techniques is to develop collaborative writing pedagogies more akin to theatrical improvisation where the traditional roles of actors and spectators are merged into more dynamic, cooperative positions. The practice might even compare to what critics and playwrights have long termed removing the fourth wall between stage and audience.

This persistent absence of this wall in digital media is probably best theorized in the anthology by Torres and Maduro in "Wreadings." Digital authoring tools in their earliest formats, like Flash and JavaScript, they note, immediately broke down any distinction between "reading" the web and "writing" or authoring it. In fact, "online reading experiences" appeared to diverge from print-based practices almost simultaneously with these same early authoring tools. On one level, the evolution of wreading seems to have embraced notions of conflict, where readers taking a more active position in a text's construction seem both liberating and progressive. The traditional role of the author as a solitary figure (usually male) shepherding readers en masse continues to lose all cultural capital in the digital age. Here again, algorithms newly deployed across the web to help distribute and produce content autonomously, free of individual aims and desires represent a supposedly liberating mechanism, establishing rules where wreaders might gain more cultural influence over digital textuality. The wreader, Torres and Côrtes Maduro contend, should not be considered the pivotal agent of digital media – especially given this agent's original incarnation as an early hypertext system operative. However, as digital media theorists strive to point out, hypertexts continue to bear a narrative structure just as linear as a print text, due to how predictable and sequentially linked they can be. The algorithms informing early hypertexts may be categorically basic, to recall Finn's classifications; yet, as algorithms become more sophisticated, so, too, do any texts they may subsequently generate, leading, in turn, to more complex wreading roles and techniques.

Not surprisingly, few, if any, of the works discussed in this collection invoke the techno-optimism of a new 21st century *Whole Earth Catalog* community. One might detect the allure of the unknown that haunted Breton's cry for imaginative play, but the stakes informing these experiments in computation and interaction seem far more complex and open to interpretation. Salter in her essay on developing pedagogy for teaching interactive forms of fiction and literature, "Maker Generation? The Uncertain Future of Students as Interactive Storytellers," qualifies the art of digital storytelling specifically as "a form of critical making focused on meaning-making . . . [where] [t]he weaving of connections is as important as the content to creating the story." As meticulous and rigorous as the technique may be, she argues, networks by their very structure offer no stable template for generating meaning. Variability is part of the form. A key feature of the DCL Singling program is likely the shared point of discovery it induces the precise moment any entered words into the application transform into MIDI sounds. No resulting pattern seems even remotely predictable, and the event, itself, seems partly comparable to an exquisite corpse poem or drawing being unfurled, fold by fold after the last line is drawn. This digital resurrection, too, seems best described as a kind of unveiling-as-convergence, as the work literally emerges before one's eyes and ears. Each result also underscores the important effect collaborative, interactive methods can produce when funneled through the right combination of digital constraints and tools, allowing one media form to transform into another. Out of such transformations, no end of different patterns and experimental aesthetic directions seems forthcoming. Collaboration as both a practice and mode of interpretation clearly figures prominently when applying computation to new art and writing practices. The algorithm may still be mysterious, but part of its mystery seems fundamental to the creative process in a digital environment.

Algorithmic processing, as any gamer will testify, involves cooperation as readily as it generates conflict. Where games depend upon competitive struggle as part of normal interactive play, the constraints guiding that interaction will inevitably encourage argument and contention in the interest of determining patterns and correlations in datasets. Here, pattern and integration do not necessarily invoke unity, so much as they describe the simple movement or flow of data, however rhythmic, from one place to another. Within this ecosystem and the corresponding relationship to technology it offers, objects and transactions acquire meaning primarily through their capacity to be transmitted rather than according to any prescribed ends or aims. While computation in theory attributes meaning to controlled tasks and rules of play, the resulting process, itself, can never be thought of as free from conflict or dispute. Disputation, accordingly, presents a fitting and necessary counterpart to computation in order to maintain balance within a single ecosystem.

Algorithms provide an engaging, if still highly mysterious, tool of integration, assimilation, and also disputation. They play a key role in computational relationships managing and instructing individual components in any system towards specific, usually predefined objectives. As entities in themselves, they project fascinating data arrangements, pushing through flow after flow of new sets of relations in an endless display of movement and combination. To have them at hand in devices strapped to our wrists, humming inside jacket pockets close to our hearts can feel instantly reassuring, psychologically as well as intellectually. Given these features, and the level of assimilation between platform and user they sponsor, it's easy to imagine the emergence very soon of a more powerful cultural idealism than the technocratic utopianism embraced by the nascent network culture of the early postwar period. While social networking in the current moment may not immediately reflect this level of personal faith, digital media platforms have managed to integrate a broad range of diverse fields and disciplines, blending the tools of data science and statistics almost seamlessly with projects in literary criticism, sound art, and poetics. All the more reason, each of the essays in this collection argue, to begin building an effective set of critical tools to help us engage, create, and dispute within a cybernetic state of unprecedented singularity.

Works Cited

Breton, André. *Manifestoes of Surrealism.* University of Michigan Press, 1969.

Colman, Dan. "The Whole Earth Catalog Online: Stewart Brand's "Bible" of the 60s Generation." *Open Culture,* 2011.

Finn, Ed. *What Algorithms Want: Imagination in the Age of Computing.* MIT Press, 2017.

Flores, Leonardo. "Third Generation Electronic Literature." *Electronic Book Review,* 2019, doi: https://doi.org/10.7273/axyj-3574.

Harvey, David. *A Brief History of Neoliberalism.* Oxford UP, 2005.

Kowalski, Robert. "Algorithm = Logic + Control." *Communications of the ACM,* vol. 22, no. 7, 1979, p. 13.

Kraynak, Janet. *Contemporary Art and the Digitization of Everyday Life.* University of California Press, 2020.

Miyazaki, Shintaro. "Algorhythmic Ecosystems Neoliberal Couplings and Their Pathogenesis 1960-Present." *Algorithmic Cultures: Essays on Meaning, Performance and New Technologies,* edited by Robert Seyfert and Jonathan Roberge, Routledge, 2016, pp. 128-139.

PredPol®. "The Predpol® Algorithm." Website. *PredPol®.* 25 February 2021 2020. https://www.predpol.com/technology.

Rosenblat, Alex. *Uberland: How Algorithms are Rewriting the Rules of Work.* University of California, 2018.

Seyfert, Robert and Jonathan Roberge. *Algorithmic Cultures: Essays on Meaning, Performance and New Technologies.* Routledge, Taylor & Francis Group, 2016.

Turner, Fred. *From Counterculture to Cyberculture: Stewart Brand, the Whole Earth Network, and the Rise of Digital Utopianism.* University of Chicago Press, 2006.

Part I.
Foundations

Chapter 1

Wreadings: Digital Poetry and Collaborative Practice

Rui Torres

University Fernando Pessoa, Portugal

Daniela Côrtes Maduro

University of Coimbra, Portugal

Abstract: The term wreading has been used in a diverse number of contexts, especially when the process of reading hypertext fictions is under analysis. However, we believe that, in some works of digital poetry, this term acquires a particular meaning which will be explored in this article. We will begin by examining the conditions in which this literary form emerged and evolved by exploring the specificity of poetic discourse, identifying several of digital poetry's antecedents and exploring a connection between art and technology. After describing the aesthetics of digital poetry, we will refer to the many uses of the term "wreading" and explore its association with collaborative writing. Finally, we will focus our attention on a particular type of wreading suggested by Pedro Barbosa's experimental work, and describe "collaborative wreading" as an artistic, research and teaching tool.

Keywords: indeterminacy, poetics, hypertext, collaborative writing, digital poetry

***

1. Poetic Discourse and Experimentalism

Before we focus on the emergence and aesthetics of digital poetry, it is important to clarify what is meant by poetic discourse. In poetry, the verse may present spatial formations which break with any linear arrangement, enabling not only spatialization of form (which corresponds to the concept of constellation introduced by French poet Stéphane Mallarmé, and later developed by concrete poets like Eugen Gomringer and Max Bense), but spatialization of content as well. In poetry, language is not transparent, but rather opaque, and, specially in the case of experimental poetry, the production of meaning is often stalled or problematized, thus focusing the reader's attention on the signifying process itself. In poetry, as Tzvetan Todorov pointed out, signs are intransitive symbols that do not refer to anything besides themselves (10). Therefore, in lyrical texts, symbols form an expressive network of similarities, and unlike with communication, ambiguity is not a burden, but a rationale. This sense of vagueness is the result of a blending of contrasts: in poetry, there is not a clear distinction between abstract and concrete, ideal and material, or general and particular (Todorov 10). Instead, poetic words remain *in-between*: they are both redundant and ambiguous.

Poetry is circumscribed only by indeterminacy, and it is this indeterminacy which makes poetry an expressive medium to speak the unutterable (Todorov 11). Poetry is then the expression of the ineffable, urging the reader to look for new semantic configurations, promoting unusual relationships between that which is represented as abstraction, and that which is represented by the concrete materiality of the media involved. One should not forget that this tension is inherent to language itself. However, it can only be fully grasped within literature and more precisely within poetry, because in our daily use of language, ambiguity needs to be abolished for reasons of communicative efficiency. Let us not forget that for many poets, language *is* poetry and, as Ralph W. Emerson rightly suggested, languages are composed of *fossilized metaphors*: "We are symbols and inhabit symbols (...) and being infatuated with the economical use of things, we do not know that they are thoughts. The poet (...) gives them a power which makes their old use forgotten, and puts eyes and a tongue in every dumb and inanimate object" (24-25). That is why a definition of poetry may perhaps correspond to Viktor Shklovsky's definition of art, which "exists to make one feel things, to make the stone stony (...), to impart the sensation of things as they are perceived and not as they are known" (12).

Print, as Walter Ong suggested, brings closure to, and promotes a commodification of the word (132). To Ong, the printed text remains bound to a fixed and permanent vehicle (135). However, even if page-bound, a text may be open to new possibilities of signification. Poetic language is mostly focused on the creation of unconventional correspondences between elements, opening up discourse to dynamic recreations of meaning. This poetics is particularly important to the experimentalist stance towards poetry. Contrary to Ong's assumption that print entraps texts in a stable surface, twentieth-century experimental poetries have creatively explored the surface of the page in order to find alternative (re)presentations of both form and content. From futurists to concretists, many poets have gone beyond the graphical conventions of the page and amplify meaning beyond what is immediately perceived. In this sense, the notion of open text suggested by Umberto Eco may be useful to describe the limitless production of meaning which characterizes experimental poetry. Experimental poetry is frequently based on the notion of open work. According to the definition provided by Umberto Eco, based on Henri Pousseur, the open work produces in the interpreter "'acts of conscious freedom' (...) and places him at the focal point of a network of limitless interrelations, among which he chooses to set up his own form" (4). To Eco then, even when inscribed on a surface, "a work of art is never really 'closed,' because even the most definitive exterior always encloses an infinity of possible 'readings'". (24)

This aspect was materialized in a word. Besides challenging the idea that texts enclose an original meaning perpetuated by print's fixity, experimental poetry initiated an expansion into new territories beyond the verbal. Therefore, the current multimodal character of digital poetry can be traced back to previous experiments with the word, image and sound represented, for instance, by a word coined by James Joyce in *Finnegans Wake* (1939)[1] and frequently used by the Noigandres group of Brazilian concrete poetry: the *verbivocovisual* (Joyce) or the *verbi-voco-visual* (Noigrandres) *verbovocovisual.*[2] In a collection of theoretical texts about concrete poetry, Augusto de Campos claimed that "concrete poetry could be defined as an irreversible and functional (...) audioptical structuration, which creates a complete and dynamic "verbivocovisual" – a term created by Joyce – entity, comprised of ductile, moldable, amalgamable words, made available to the poem" (34 [our translation]). In the same book, "verbivocovisual" is considered a "linguistic field" which makes use of the many advantages of non-verbal communication without giving up on "word's virtualities" (157, [our translation]). As Philadelpho Menezes has suggested, the three basic matrixes (the verbal, the visual, and the sonorousaural) can be found in all languages. However, poetry was assumed to be dependent on the verbal sign matrix only. With the birth of modern art in the late nineteenth century, this idea started being dismantled (Menezes n. pag.). As we will demonstrate in the next section, digital media further contributed to the erosion of this notion of poetry as an exclusively

---

[1] "Up to this curkscraw bind an admirable verbivocovisual presentment of the worldrenownced Caerholme Event has been being given by The Irish Race and World" (*Finnegans Wake.* Wordsworth Classics, 2012, p. 341).

[2] This term was also used by Marshall McLuhan. See *Verbi-Voco-Visual Explorations* (1967).

verbal form by combining in an articulate and intuitive manner these three matrixes. Consequently, digital media epitomize the collapse of any divisions between art forms.

2. Art and Technology: the emergence of digital poetry

Digital media introduced exciting possibilities for experimentation by bringing into play opacity, ambiguity, indeterminacy and intermedia experiences. Hypertext and hypermedia unlocked further opportunities for artistic expression, while at the same time providing authors with critical tools to challenge preconceived notions of author, text and reader. New media theorists agreed that digital media was the result of a fusion between visual culture (photography, film, TV) and computing technologies.[3] As a matter of fact, by resorting to digital optical storage, digital computer allowed artists to combine, for the first time, some of the most emblematic features of film, photography, TV and telecommunications. This interconnection of art and technology ultimately led to the emergence of digital poetry. There seem to be four key moments in the twentieth and twentieth-first centuries which strongly contributed to the development of digital poetry (in fact, though we decided to focus here on recent manifestations, some of digital poetry's particular features can be traced as far back as the Baroque period). This tentative timeline, therefore, does not represent an attempt to circumscribe the ever-evolving field of digital poetry, but aims to emphasize an interconnectedness between experimental poetics and technological infrastructures.

First, in the beginning of the twentieth century, the film established the theory and the practice of editing and montage techniques, articulating a cultural tendency for *bricolage* with the artistic concept of collage pioneered by the typographic experiments of futurists and dadaists. Artists such as Tristan Tzara, F. T. Marinetti and Kurt Schwitters, pursued montage and juxtaposition as a means for the production of unrelenting and fluctuating meaning, a process that faithfully characterizes digital poetry.

Later in the middle of the century the development of TV and digital computers encouraged the outbreak of experimental poetries. Concrete, visual and sound poetry of the 60s claimed new horizons for poetry, and groups like the PO.EX (Portugal), Noigandres (Brazil), and L=A=N=G=U=A=G=E (U.S.A.), as well as several experiments with combinatory and automatic texts generated by computers—such as Nanni Ballestrini in Italy, Pedro Barbosa in Portugal, and

---

[3] See Bob Cotton and Richard Oliver, *Understanding Hypermedia* (London, Phaidon, 1993), and Lev Manovich, *The Language of New Media* (Cambridge, The MIT Press, 2001).

A.L.A.M.O. (Atelier de Littérature Assistée par la Mathématique et les Ordinateurs [Literature Workshop Assisted by Mathematics and Computers]) and L.A.I.R.E. (Lecture, Art, Innovation, Recherche, Écriture [Reading, Art, Innovation, Research, Writing]) in France—became particularly relevant for the development of digital poetry. Urged by the arrival of the computer, art and technology were brought together, and experimental poets of many countries started remixing materials made available by recording technologies, while creating postmodern re-readings of established or published texts. For instance, E. M. de Melo e Castro (PO.EX), pioneering writer of concrete poetry and videopoetry in Portugal, resorted to the Portuguese national television studios and to new digital and electronic equipment (Reis 105) to endow printed letters with movement. This experiment resulted in the creation of *Signagens* (1985-1989), a series of videopoems comprised of "a combination of letters, symbols, and visual effects" (Funkhouser 123).[4] In a third stage, as culture became networked and represented by the metaphor of the rhizome,[5] Internet challenged ingrained notions of community and space. It is precisely this cyberculture (culture developing in the cyberspace) which shaped the emergence of digital poetry by suggesting alternative montage and collage techniques, associating *wreading* with collaboration (a trend which will be analysed in the final section of this article), and merging multimedia with programming. As C. T. Funkhouser emphasized, "most of the mechanical groundwork for digital poetry was already established by the time the WWW became prominent" (7). Funkhouser identified the existence of a "textual and graphical cannibalism" which he considered digital poetry's "strongest trait" (239). However, thanks to new authoring tools (for instance, Flash and JavaScript), additional textual behaviors were introduced. Readers could now explore online texts comprised of sound, moving image and opportunities for interaction. Antero de Alda's *Scriptpoemas*[6] are a remarkable example of this multimodal expansion of digital poetry. In this work, Alda invites the reader to activate and manipulate seventy-five poems comprised of animations, sound and videos, thus challenging the traditional notion of a poem as fundamentally comprised of words and stanzas.

The fourth and last moment is deeply connected with the emergence of augmented and virtual reality, as well as with the use of mobile devices to read

---

[4] For the concept of network, see Manuel Castells, *The Rise of the Network Society* (Oxford, Blackwell Publishing, 1996).

[5] For the notion of rhizome see Gilles Deleuze and Félix Guattari, *Mille plateaux* (Paris, Éditions de minuit, 1980).

[6] 2005, www.anterodealda.com/scriptpoemas.htm. Accessed 26 May 2020.

and create digital poetry. Due to mobile devices, as well as tracking technology, readers of digital poetry are no longer limited to desktop screens or to mouse clicking in order to access and explore a text. Technology allows, once again, the emergence of new textual behaviors and, besides clicking or scrolling up and down, more user actions are implicated in the reading process. For instance, while reading María Mencía's *Transient Self-Portrait*,[7] a re-reading of "En tanto que de rosa y azucena" (1522) by Garcilaso de La Vega and "Mientras por competir con tu cabello" (1582) by Luís de Gongora, a portrait of readers facing the webcam becomes part of the poem and, besides mouse actions, their voices are used to activate the text. This work is the result of a collaboration between María Mencía (concept and research), Reda Yacouby and Jalal Miftah (programming), Alix Ingber (translation to French), Pierre Darmangeat (translation to English), Pablo Romero (voice - Spanish) and Serge Bouchardon (voice - French).

All the above-mentioned poetries, whether futurist, concrete, or digital, point to particular ways of interpreting (as well as recreating) the world by making use of the barely explored affordances of new technologies. In the next section, we will inquire how the use of technology contributes to the construction of an aesthetics exclusive to digital poetry.

3. The Aesthetics of Digital Poetry

Many agree that digital poetry builds on the achievements of experimental poetics and that visual, sound and concrete poetries are developed and renewed by e-poetries. However, reproductions of print-based poems that were made available online after the popularization of the WWW cannot be considered examples of digital poetry. As theorists from different fields of study (for instance, literature or media studies) have argued, it is possible to identify two main tendencies emerging in the beginning of the twenty-first century.[8] The first implies a *replication* of static text onto the medium of the web, and the second, as Kenneth Goldsmith explains, "engages directly with technology to provide (...) a new type of online reading experience (...)" (i). As Loss Pequeño Glazier pointed out, despite the obvious connections to previous avant-garde experiments, digital poems should not be considered an extension (digitisation, or remediation) of the printed form, but a particular way of thinking about and through the media instead: "the idea of the digital poem as the process of thinking through this new medium, thinking through making" (6). When Glazier

---

[7] 2012. www.mariamencia.com/pages/transientself_portrait.html. Accessed 26 May 2020.

[8] For the analysis of remediation as a general trend of new media, see Jay David Bolter and Richard Grusin, *Remediation: Understanding New Media* (Cambridge, MIT Press, 1999).

declared that "the web we read is the web we write on," he considered poems as both the result and the process of creative practice (33).

While experimenting with the medium in an attempt to know its potentialities and limitations, poets paved the way for the creation of digital poetry. Reframing Emerson's expression, one may say that digital poetry "defossilises" the technological language, enlightening us about the medium itself. By avoiding transparency and fixity, digital poetry proposes a self-reflexive stance, calling our attention to the materiality and the ontology of the medium. Like experimental poetry, digital poetry presents its intransitive symbols through self-reference, and its main features could be described, either specifically or symbolically, in the following manner: metaphorical infinitude (randomness and open-endedness); *wreading* (re-writing and *plagiotropia*); multimodality (integration and convergence); customization (manipulation and selection) and networked creativity (open-access, instantaneous and self-publication, and collaboration).

4. Wreading and Collaborative Practice

Though running the risk of excluding writers who, located in other countries and making use of different languages, may have also resorted to the term *wreading,* we believe that it is important to explore the different uses of this word (and by extension, of the word *wreader*) to fully examine the nature of collaborative practice in digital poetry. In this last section, we aim to highlight four different uses of the term *wreading* and understand how this word came to be associated with "collaborative writing."

The term *wreading* is the result of the fusion of two different words: writing and reading. In the early nineties, when the term *wreading* is believed to have emerged (or at least, when it became widespread), it aimed to represent the increasingly active role of the reader during the contact with a text, specially a digital text. Interestingly, it was introduced in different languages, and even though employed in various ways, the emergence of this term was, in many cases, simultaneous. Joe Amato, in a review of J. David Bolter's *Writing Space: The Computer, Hypertext, and the History of Writing* (1991), suggested the English word *wreader* to describe the "reader-cum-writer" concept introduced by Bolter. Amato related this term with a text that allows readers "to make (finite) choices about what to read next even as they proceed through interactive texts, choices that control the sequencing of the text itself" (li. 75). The term was later used in the context of hypertext theory and digital literature by George P. Landow who claimed that "in hypertext the function of reader merges with that of author and the division between the two is blurred," and that "[t]echnology transforms readers into reader-authors or 'wreaders'" (*Hyper/Text/Theory* 14).

In 2000, Arnaud Gillot published a book titled *La notion d'«Ecrilecture» à travers les revues de poésie électronique* alire *et* KAOS [The notion of «Wreading» through electronic poetry journals *alire* and *KAOS*]. In the same year, Mark Amerika associated the term *wreading* with "all of the hyperrhetoric around hypertextual narrative" (par. 7) digital art and "some of the more experimental postmodern metafiction of the post-War period" (par. 8). To Amerika, hypertext narrative was "'a revolutionary mode of publication' that changes readers into 'wreaders'" (par. 8). He also compared the *wreader* to an "interactive co-participant" or the "co-conspirator" (par. 8) of Julio Cortazar's "prototypical postmodern narrative" of *Hopscotch* (1963), thus emphasizing the widely accepted link between hypertext fiction and postmodern literature. Even though this term was well-established in 2000, it did not receive unanimous approval. Nearly a decade before Amerika, in 1992, Martin Rosenberg had related the "capacity of 'wreaders' to jump through links from lexia to lexia, forwards, backwards" with a "control over experience" which "reminds one of the control over the natural environment enabled by the calculus of Leibnitz and Newton" (Rosenberg, "Contingency, Liberation and the Seduction of Geometry" n. pag.) Rosenberg believed that, although hypertext theorists related the term *non-linear* with recursiveness and chaos, hypertext systems were in fact fostering a "rhetorics entrapped in the necessarily logocentric geometry of regulated time and space" (ibid.). Hypertext proponents argued that the "contingencies of choice" liberated *wreaders* from the "linearity of narrative, or from the linear mono-logic of academic argumentation" (idem.). However, to Rosenberg, since the *wreader* was unable to controvert hypertext's logic, and was rather urged to believe in its "functional transcendence," this system of choice produced an effect opposite to that intended: to Rosenberg, *wreaders* were compelled to "assume naively" that they had the freedom to explore and establish their own territory. Consequently, Rosenberg believed that hypertext was "inadequate as a vehicle for that [avant-garde] transgressive and liberating moment" because it failed to sustain the dislocation experienced in the jump between lexias. To Rosenberg, hypertext replaced narrative linearity with predictable and sequential linking[9]: as soon as the leap between lexias was completed, "a normalcy emerged" and the "space-time black-hole adventurer, or the hypertext reader" became easily acclimated to this fragmentary reading.

---

[9] In 1997, Espen Aarseth made a similar observation by stating that hypertext "can be a much stronger linear medium than the codex, should its author decide so," and "a hypertext path with only one (unidirectional) link between text chunks is much more authoritarian and limiting than (say) a detective novel, in which the reader is free to read the ending at any time" (46-47).

In 2000, Anja Rau also attempted to demystify the supposedly participatory role of the reader in the co-creation of the text. Rau observed that, instead of shaping the text "after his or her needs and desires," readers of hypertexts, and more specifically hyperfictions, "often complain[ed] that they [felt] at the mercy of the text, pulled hither and thither by invisible paths and hidden guard fields." In fact, Rau identified a point of diversion in hypertext theory, and added that "second generation hypertext-critics even claim[ed] that hypertext imposes far more restrictions on the text and the reader than good, old linear story-writing."

In a book dedicated to the pair "immersion and interactivity," Ryan described the word "wreader" as "a self-serving metaphor that presents hypertext as a magic elixir: 'Read me, and you will receive the gift of literary creativity'" (9). To Ryan, interactivity had been used by hypertext theory as a tool to disempower the author. However, as many other authors, Ryan argued that the process of writing could not be equated with the syncopated itinerary suggested by Landow's hypertext.

Jim Andrews also refused to see *wreaders* as a group of overpowering readers who finally gain control of the text. Andrews uses the term *wreader* to describe the users of two of his texts, *Nio*[10] and *Stir Fry Texts*,[11] both published in the first volume of the *Electronic Literature Collection*. Nevertheless, to this author,

> the 'wreader' does not compose texts from scratch with the stir frys, but instead explores/composes with 'content' that is already within the piece. This is true of *Nio* and the other interactive audio work I've done so far, though it is easy to see that both could be developed as tools alone. But I am not drawn to making tools alone. (Andrews, "Nio & the Art of Interactive Audio for the Web" n. pag.)

Therefore, in contrast to the belief shared by some theorists who used the term *wreading* in its embryonic stage, Andrews rejects the idea that hypertext readers are given the necessary tools to co-create the text together with the author. Quite the contrary, while equating the process of reading the text with the manipulation of content pre-determined by the author, Andrews denies the reader the ability to pursue such a goal.

---

[10] Jim Andrews, *Nio* [2001]. *Electronic Literature Collection*, vol. 1, 2006. collection.eliterature.org/1/works/andrews__nio.html. Accessed 26 May 2020.

[11] Jim Andrews, *Stir Fry Texts* [1999]. *Electronic Literature Collection*, vol. 1, 2006. collection.eliterature.org/1/works/andrews__stir_fry_texts.html. Accessed 26 May 2020.

In 2003, Michael Allen argued that, in a *wreader*, "both production and consumption of texts is combined into one process that is self-contained." (Allen, "This Is Not a Hypertext, But..." n. pag.) He compared George Landow's *wreader* to Jerome McGann's radial readers who "read texts in an open-ended search for meaning" (Ibid.). At the time, Allen also mentioned the possibility that "[n]ew hypertexts might adapt to Internet technology's image-based meaning systems and render Landow's new reader archaic." (Ibid.) In fact, as we have already remarked, web's multimodal tools made available to authors effected profound changes in digital literature. After the emergence of the WWW, works were no longer limited to the image of a standalone object created by a single author, as was the case of Storyspace hyperfictions. They could now assume the shape of digital artefacts accessed online and created by several contributors.

As we have seen, the word *wreading* is often related to freedom of choice awarded to the reader, in an attempt to overturn the authorial regime. This insurgency against the author is particularly illustrated by hyperfictions such as Michael Joyce's *afternoon: a story* (1987-1990), Stuart Moulthrop's *Victory Garden* (1992) or Shelley Jackson's *Patchwork Girl* (1995). Based on a kinship—envisioned by George Landow or Jay David Bolter, among others—between hypertext theory and poststructuralist theory, these multilinear works consisted of text chunks interconnected by links which could be traversed by the reader. As such, we will name this process *hypertext/post-structuralist wreading*. However, since some works, even though not advocating post-structuralist or hypertext theory, may also create the illusion that readers have an increased participation in the construction of the text, we would like to acknowledge a second type of *wreading*: the *metaphorical wreading*. This type of *wreading* is not associated with an attempt to undermine authorial power, but to a creative exploration of the aesthetic potential of interactivity. In these works, readers perform several configurative and assembly tasks to reveal parts of the text, and this process may grant the impression that the reader actively collaborates in its writing. However, as emphasized by many of the above-mentioned theorists, the reader is merely following the protocol established by the author/ programmer. This procedure is illustrated by, for example, the reading of *The Sweet Old Etcetera*, an interpretation (or as we will later explain, *appropriative wreading*) of e.e. cummings' poems created by Alison Clifford.[12] In this work, readers are requested to activate sounds and

---

[12] Alison Clifford, *The Sweet Old Etcetera* [2006]. Electronic Literature Collection, vol. 2, 2011. collection.eliterature.org/2/works/clifford_the_sweet_old_etcetera.html. Accessed 26 May 2020.

verses in order to keep reading, thus matching the process of revealing the text with the process of writing the text.

All the theorists mentioned so far relate the process of *wreading* to the sum of interchanges occurring in a closed circuit between the reader and the digital text. Nevertheless, many authors have explored other applications of the word *wreading*. In 2002, Jed Rasula applied this same term to print literature by arguing that *his* neologism, *wreading*, represented a "collaborative momentum initiated by certain texts (...) in which the reader is enlisted as an agent of the writing.[13] Reciprocally, the writer discloses his or her own readerly orientation (albeit in ways that vary from work to work)" (11-12).

In 1991, the same year Joe Amato published his review of George Landow's *Writing Space*, Pedro Barbosa coined the Portuguese word *escrileitura* (*escrita* [writing] and *leitura* [reading]).[14] However, though also commenting on the use of the computer as a tool for literary creation, he suggested a further meaning of this word. In his *Teoria do homem sentado* [Theory of the Sitting Man], Barbosa explains the mechanism of *Syntext*, a program, or "textual generator," created during his experiments with computers conducted in the seventies. According to Barbosa, while using this program, texts (or more precisely, virtual texts) exist in a "potential state" and the computer becomes a "complexity telescope." Readers are expected to use the computer or, in Barbosa's words, "the actualizer of the text's potential" by "reading, selecting, correcting, adding, changing and recording the large amount of material that the textual generator made available in a fraction of a second" (10). In this process, the user ceases to be the reader, and is turned into a "co-writer" (40) or "wreader", who "practices reading as writing, and writing as reading, in a new interactive symbiosis" (Barbosa 11). This work by Barbosa was repurposed by Rui Torres using his own poemario.js, programmed by Nuno F. Ferreira in 2016, and the resulting version was published in the *Electronic Literature Collection*.[15] The lib poemario.js used to emulate Pedro Barbosa's

---

[13] In the comments to an entry published in the blog *Now What, a collective blog by alternative prose writers & publishers*, co-founded by Lance Olsen and Ted Pelton, Joe Amato claims that Jed Rasula coined the term in the early 1980s while describing "a certain kind of poetry/poetic experience." To read this discussion, see: nowwhatblog.blogspot.com/2006/06/matt-madden-99-ways-to-tell-story.html. Accessed 26 May 2020.

[14] Barbosa used this word in his PhD thesis in 1991, *A Ciberliteratura: Criação Literária e Computador*, later published in 1996 (Lisbon, Cosmos).

[15] Pedro Barbosa, *Cyberliterature (1977-1993)*. Electronic Literature Collection, vol. 3, 2016. collection.eliterature.org/3/work.html?work=cyberliterature. Accessed 26 May 2020.

work was made available by Rui Torres and Nuno F. Ferreira[16] so that all users could create their own *wreading* of any text. Therefore, Pedro Barbosa's *Syntext* and poemario.js suggest a third type of *wreading.* Instead of allowing the reader to choose between bifurcating paths, Pedro Barbosa offers *wreaders* a tool that enables them to shed new light on a source text by re-writing it indefinitely. Thus, the kind of *wreading* (here referred to as *appropriative wreading*)[17] suggested by Pedro Barbosa involves the manipulation of a source text which, thanks to the computer, inhabits a different plane of existence (a "virtual" existence, according to Barbosa), a condition which enables the reader to continuously reframe or amplify the meaning(s) of a source text.

Simanowski also uses the term *wreader* in a text published in 2001. However, he acknowledges the elusive character of this term, and refuses to proclaim the death of the author, announcing the death of the reader instead: "The author is to set up the links and therefore to control the reader's associations. And since the author—not the reader—knows the password for the File Transfer Protocol she still controls the text after it has been published" (Simanowski, "Death of the author?" n. pag.). Simanowski replaces the term *wreader* with "restless traveller through the world of text" or with "viewer" (ibid.). As for the possibility of the reader to exchange places with an author, this possibility is not, according to Simanowski, associated with the process of *wreading,* but with collaborative writing per se. As stated before, *wreading* is often related to a collaboration between reader and author in the writing of the text, but Simanowski prefers to highlight the gap between writing and reading: "[w]hat I would like to talk about is the reader's transformation into an author. This does not refer to the wreader concept that has been applied to hypertext. This refers to collaborative writings, where readers *are invited* to become authors." (ibid. [our italic])

Many online platforms not intended for literary creation (Facebook or Twitter, to name but a few) host collaborative writing projects in the sense described by Simanowski. One of the first platforms being used for online collaborative writing was the WELL (Whole Earth'Lectronic Link). In fact, the creation of Judy Malloy's *Uncle Roger* (1986-), often considered as the first work of digital literature, resulted from an invitation made by Carl Loeffler to

---

16 Instructions to use *Poemário* are available (in Portuguese) online. telepoesis.net/poemario/?page_id=2. Accessed 26 May 2020.

17 This kind of *wreading* may exist in printed form. However, in this article we will focus mainly on digital wreading processes.

participate in Art Com Electronic Network (ACEN) and to write on the WELL. This platform, founded by Stewart Brand and Larry Brilliant in 1985, was described by Howard Rheingold, in *The Virtual Community*, as "a computer conferencing system that enables people around the world to carry on public conversations and exchange private electronic mail" (1). However, this system was not a mere platform for communication. In this excerpt, Rheingold describes the improvisational and collaborative nature of the WELL and mentions the existence of an interlocking of roles which reminds us of the term *wreading*:

> [f]inding the WELL was like discovering a cozy little world that had been flourishing without me, hidden within the walls of my house; an entire cast of characters welcomed me to the troupe with great merriment as soon as I found the secret door. Like others who fell into the WELL, I soon discovered that I was audience, performer, and scriptwriter, along with my companions, in an ongoing improvisation. A full-scale subculture was growing on the other side of my telephone jack, and they invited me to help create something new. (2)

Rheingold describes the improvisational and collaborative nature of the WELL and mentions the existence of an interlocking of roles which reminds us of the term *wreading*.

In a paper presented at one of 10th ACM Conference on Hypertext and Hypermedia workshops, in 1999, Heiko Idensen reframed the role of the reader by conflating reading with writing. According to Idensen, readers and writers "write and read simultaneously at one distributed and dismembered texture, that is spread over the whole world" (n. pag.). Instead of linear writing, Idensen suggested a "poetics of link" or "poetics of transportation" that encouraged "a nomadic action of roaming around in media networks" and a "collective TEXTUALITY" in which "reading and writing will overlap" (ibid.). "Kollaboration" was thus linked to an attempt to dismantle systems of control or, according to Idensen, "act against totalitarism" and "mono-culture" by paving the way to a "new collaborative discourse-technics" (ibid). Rhizomatic and networked reading/ writing was described by Idensen as a collective and transgressive practice which is also used to "eliminate bourgeois author-subject" (ibid).

In a more recent article, Isabell Klaiber argued that participants in collaborative online writing projects developed at platforms such as One Million Monkeys, Protagonize, StoryPassers, Ficly, or WEBook (124) may "truly be considered a 'reader-author' or a 'wreader'" (130) because, in order to participate (or become a writer, and exchange roles with the author) in the writing of a collaborative narrative, they have to read texts which were previously published in the same platform. Therefore, as Klaiber demonstrates, *wreading* has been also used in a broader sense (we will name this sense *networked wreading*) to describe the activity of those that write/ read/ publish/ edit texts in blogs, forums and platforms intended (or not) for collaborative writing. In fact, M. José Luzón and M. Noelia Ruiz-Madrid mention the existence of a "wreading competence" or the "ability to read online texts and construct one's own texts" (1). To these researchers, "wreading competence" involves the following requirements:

> understand the pragmatic, discursive and semiotic features of online texts, harness their affordances and interact with them in various ways, find relevant information in different semiotic modes within and across these texts, and relate and meaningfully use such information in order to achieve a specific purpose, complete a task or produce an output. (Luzón and Ruiz-Madrid 1)

As we have seen, digital media allow the modification of texts, decisively eroding the notion of author(ity) and text as permanent and fixed entities. While providing the opportunity to work in a multimodal and networked environment, digital media encourage what Roy Ascott once named "distributed authorship" in *La Plissure du Texte.*[18] Digital poetry takes advantage of the medium's affordances, involving sometimes whole teams of researchers and artists, as well as different skills (from programming to illustration), to create a single work. Scott Rettberg identified three possible degrees of participation in digital collective literary and artistic production which may be used to contrast *collaborative writing* and *wreading,* as suggested by Simanowski (see above). In *contributory participation,* "contributors may not be aware of how their contribution fits into the overall architecture of the project, or even of the nature of the project itself, but they do take conscious steps to make their contribution available to the project" (Rettberg n. pag.). As for the *unwitting participation,* "texts utilized in the collective narrative are gathered by the text-machine itself, and contributors have no conscious involvement in the process of gathering the material" (ibid.). In *conscious participation,* however, "contributors are fully conscious of explicit constraints, of the nature of the project, and of how their contribution to it might be utilized" (ibid.). Therefore, in *conscious participation,* contributors are aware of their role and of the extent

---

[18] Roy Ascott, *LA PLISSURE DU TEXTE: a distributed authorship project for ARTEX, december 11-23, 1983.*

and nature of their participation, since they are indeed, to use Simanowski's words, "invited to become authors". This kind of participation, the one we would like to explore in this article, is not available to readers performing *hypertext/poststructuralist* and *metaphorical wreadings*. As *wreaders* need to be aware of the goals of a certain project and know how they can contribute in order to "consciously" participate in it, this mode of participation is exclusively available to those engaged in collaborative projects comprised of several members (and not just "the reader" and "the author"). As we hope to demonstrate, *networked* and *appropriative wreadings* allow a collaboration between several members, and therefore, contrary to *hypertext/poststructuralist* and *metaphorical wreadings*, these types of wreadings allow a *conscious participation*. In line with Simanowski, we also see the processes of ew*wreading* and *collaborative writing* as fundamentally different, and since the relationship between both has been irrevocably associated with the belief that hypertext readers could exchange roles with the writer — according to George Landow, "[w]ithin a hypertext environment all writing becomes collaborative writing" (Landow, "Hypertext" 88), we would like here to dedicate our attention to a particular notion of the term *wreading*. In this article, the process of *wreading* is dependent on a process of re-writing, whether performed collaboratively or individually. Therefore, following in Pedro Barbosa's and Simanowski's footsteps, we will focus primarily on "conscious" collaborative practices which involve a *wreading* (or reframing) activity of an already existent text, be it printed or digitally composed. Our next example demonstrates how *appropriative wreading* can be enmeshed with *networked wreading* and how both, contrary to *hypertext/postructuralist wreading* and *metaphorical wreading*, are compatible with "conscious" collaborative practice. A project of collaborative writing on the internet created by Johannes Auer, together with Reinhard Döhl, Beat Suter, René Bauer, Martina Kieninger and Sylvia Egger, has the suggestive name of "The Famous Sound of Absolute Wreaders" (2003). In this project, "six authors/readers" work on six works of digital literature to create new texts which are then "edited by a human editor and a machine" (Suter and The Absolute Wreaders n. pag.) and sampled to a radio show ("Kunstradio" of the Austrian broadcasting company ORF) by two actors/*wreaders*. The six texts become the source texts for six new works "of which at least three deal with visualization of text and metatext in the net-environment" (ibid.). Therefore, by making use of the tools provided by the web, the "absolute wreaders" turn texts into a set of citations. Auer's 'concrete_maschine' is used to free language "of the ruling code" by transforming texts into Mondrian-like pictures. Therefore, Auer, Döhl, Suter, Bauer, Kieninger and Egger are members of a collaborative and multimedial project and "consciously" participate in collaborative *networked* and *appropriative wreadings*. The "concrete_maschine" was made available for free use and, according to these "wreaders", is capable of translating "anyone's texts into images". Users of the "concrete_maschine" can submit their own texts, and thus, have a *contributory participation* in Jonathan Auer's project.

According to Manuel Portela, while reading Rui Torres' works (or "textual instruments"), the "rearrangement determined by readers' choices (...) make readers experience the productivity of the original structures and of a given author's lexicon" (44). Therefore, readers are given the chance to take a glimpse at the authoring process and use the text to perform their own *metaphorical wreadings* (rather than *wreading*, Manuel Portela mentions the existence of a "meta-reading" or a reading which is aware of its "strategies and protocols" (64). In a similar fashion to hypertext readers endowed with the freedom of choice between paths performing their *hypertext/post-structuralist wreadings*, readers allowed to manipulate or customize an already existent text are not provided with adequate tools to collaborate with the author in the construction of the text. Therefore, while Torres is able to perform a conscious *appropriative wreading*, *wreaders* of his works are agents of a *metaphorical wreading*, because they are introduced to a specific multimodal combinatory work created as a result of a previous collaborative project[19] comprised of several contributors (henceforth named "telepoesis.net collective") invited to perform different tasks. However, some of Torres' *wreadings* allow readers to save iterations, or their own *appropriative wreadings* of the work chosen by Rui Torres. These *wreaders* have a"contributory participation" in his project, but do not collaborate with Rui Torres in the construction of his *appropriative wreading* of the text: this was already accomplished during his collaboration with invited researchers and artists. Furthermore, contrary to "concrete_maschine", *Syntext* and *poemario.js*, which allow readers to perform *appropriative wreadings* of any work, these iterations are focused on a single work which was already repurposed by Rui Torres and his team.[20]

In the *appropriative wreadings* conducted by the telepoesis.net collective, texts are turned into hypertexts. This concept, however, is not used in the sense of the interconnected database envisioned by Ted Nelson, but rather in

---

[19] Due to the theme addressed in this article, we have selected works that involve participation of actors, programmers and designers. However, Torres also performs individual *wreadings* of works. All of Torres' poems can be accessed here: https://telepoesis.net/poesias.html.

[20] See Rui Torres, "Do texto digital ao cibertexto: Percursos de escrileitura." *Entre Margens e Centros: Textos e Práticas das Novas Interculturas*. Porto, Edições Afrontamento, 2013. bdigital.ufp.pt/bitstream/10284/6604/1/rtorres_CEI_artigo.pdf. Accessed 26 May 2020.

the sense suggested by literary theorist Gérard Genette, who wrote about intertexts, metatexts, and other forms of paratextuality. For Genette, in the process of quoting, adapting, or re-writing a text, the antecedent, or the imitated literary text, is a "hypotext", whereas a text imitating a hypotext is defined as a "hypertext." In addition, a hypertext becomes "transvaluative" when it is the object of *aemulatio*, "when it not only articulates values different from those of its targeted hypotext but also substitutes its values for those in its antecedent" (Genette 2). For instance, *Amor de Clarice* is based on words, sentences, metaphors and images from the short story "Amor" by Brazilian writer Clarice Lispector. The poem is the hypertext, and Lispector's story is the hypotext.

After writing the hypertext in printed form, Rui Torres imported it to Macromedia's Flash program, adding image, sound, animation, and interactivity. In the Flash movies, the hypotext is also present as a multi-layered and visually appealing, though static, background. Lispector's hypotext can also appear in the foreground, but only as a result of a search algorithm which has to be triggered by readers when they choose to click on or drag any verse of the poem. As a consequence, Lispector's hypotext becomes a (new) hypertext, and the short story is *mapped* by the poem. On the other hand, the poem/hypertext is fully visible in the foreground of the Flash movies, as well as animated, moving around the screen and following the mouse's/reader's trails. This text is draggable, includes dynamic colors, several levels of transparency and several sounds.

"Amor de Clarice" represents an open work which builds up on the notion of shared authorship. Besides being comprised of the words selected by Clarice Lispector in the hypotext, or the short story "Amor," this work depended on a "conscious" collaboration between Rui Torres (programming, animation and texts), Carlos Morgado and Luís Aly (music and sound), Ana Carvalho (video), and Nuno M Cardoso (voice). Music was not copyrighted, and the source code of the Flash movies was made available online. In 2008, a new version of this poem was programmed, expanding the hybrid process of writing, textual combinatorics and variability of the previous version. "Amor de Clarice v.2" develops into a set of combinatorial poems that allow (w)readers to change the very paradigms that feed the original syntax. Its lexicon is comprised of several novels by Clarice Lispector. The original soundtrack was also fragmented and organized into databases that play randomly and, in addition to changing the poems, (w)readers can automatically publish their versions/*appropriative wreadings* of this work on a weblog.

The work of the following writers was the object of telepoesis.net collective's *wreadings*: Abílio-José Santos (*Releituras de obras de Abílio-José Santos*, 2019, with Bruno Ministro and Diogo Marques), António Gedeão, António Ramos

Rosa, Fernando Pessoa, Herberto Helder, Miguel Torga and Ruy Belo (*Árvore*, 2018), António Ramos Rosa (*Estou Vivo e Escrevo Sol*, 2016), Ana Hatherly (*Fantasia breve, a palavra-espuma*, 2016), José-Alberto Marques (*Gerador de Homeóstatos*, 2015), E. M. de Melo e Castro (*Álea e Vazio*, 2014), Salette Tavares (*8 brincadeiras para Salette Tavares*, 2010, and *Cantiga*, 2012), Edgar Allan Poe (*1 corvo nunca mais*, 2009), Fernando Pessoa and Sophia de Mello Breyner Andersen (*Do Peso e da Leveza*, 2009), Herberto Helder and Raul Brandão (*Húmus Poema Contínuo*, 1999-2008), António Aragão (*Poemas encontrados*, 2007), Florbela Espanca (*Amor-mundo ou a vida esse sonho triste*, 2006), Sophia de Mello Breyner Andersen and Lewis Carroll (*Mar de Sophia*, 2005).

Torres usually works with Luís Carlos Petry (pictures), Luís Aly (sound), Nuno M Cardoso (voice) and Nuno Ferreira (programming), but he also invites other actors, artists, musicians, researchers and programmers to perform his *appropriative wreadings*. Many of telepoesis.net collective's works are collaborative *appropriative wreadings* in which different skills are mobilized to dismantle the apparently stable and immutable surfaces of works. Therefore, many of the works created by this group cover the two meanings of "collaboration" suggested by Thomas Jensen Hines (4). Besides establishing a collaboration between artists, the telepoesis.net collective also creates a *collaboration of the arts* by combining two or more works (or *hypotext* and *hypertext*) in a single work. In this process, works are reshaped as multimodal entities and represented as living organisms. However, digital media are not responsible for endowing texts with life: while taking advantage of the recombinant potential of language, they merely work as magnifying glasses that highlight some of their intrinsic features.

Manuel Portela argued that "[m]ost of Rui Torres' computer works may be said to function, at the same time, as a new work and as a critical analysis of his source texts" (43). Besides using the process of *wreading* to critically address literary works, Rui Torres also uses his collaborative *wreadings* as pedagogical tools. Since digital media imply the use of several cognitive modes, they can cover multiple disciplines, from arts to programming. In fact, because digital media allow access to different types of documents, sources of information and editing tools, they encourage critical and associative thinking, as well as creative and collaborative experimentation. Thus, learning becomes a multimodal and interdisciplinary experience open to different fields of knowledge and traditional or emerging pratices.

In addition to promoting digital literacy, the process of *appropriative wreading* also involves the revaluation of traditional notions of reading and writing. A pedagogy of literary texts, as explained by Marcel Cornis-Pope and Ann Woodlief, currently implies a model of literary interpretation that must be based on "strategies of rereading/rewriting as part of a community of readers"

(147). The aforementioned digital poems created by the telepoesis.net collective, therefore, entail a pedagogy that is both critical and interactive, offering an opportunity to move "from reading to writing, and from understanding to reformulation" (Cornis-Pope and Woodlief 160). Inside the classroom, collaborative and *appropriative wreadings* also encourage the integration of criticism in the creative process, promoting insightful intertextual links between different texts that are part of the canon, and promoting interactions between writing and coding. The number of possible variants that combinatory poems created by the telepoesis.net collective generate is so enormous that the playful and aesthetic points of view collide, fostering the critical analysis of semiotic structures present in texts. This form of literary education demonstrates the intertextuality immanent to literary language, but also depicts writing as an interdisciplinary practice, integrating the art of writing and computer programming, exploring different cognitive modalities and stimulating aesthetic enjoyment. In fact, *poemario.js* was created with the intention of expanding metaphorical infinitude, *wreading*, multimodality, customization and networked creativity, and this is somehow echoed in its declared ambition: "Read / Listen / Combine: By clicking on the words, the reader can combine them from lists created for the purpose; (...) Contribute / Share / Conserve: In addition to changing the poems, the reader can save their versions / readings on a blog."

The creative and collaborative use of networked digital media in a classroom context is being promoted by a vast number of artists and researchers. In a course titled "Networked Narratives," Mia Zamora addressed the question of how media consumers are turned into media producers" (151). Though not focused on digital poetry, the course exemplifies how digital media can be used to foster collaborative practice in a classroom context by performing *networked wreadings*. Zamora's course is focused on "the negative effects of the digital mediascape" and aims to create "more nuanced (and empowered) conversations about what is possible when contributing and participating in digital networks (...) in the age of amplification and post-truth politics" (152). "Networked Narratives" (or #NetNarr), is a "networked 'collaboratory' in digital storytelling, world-building and co-learning," and its "open community of storytellers" gather online to "inhabit personas, share and remix digital media, and explore the role of multiple identities in networked spaces" (153). Instead of co-writing or co-authoring, Zamora uses the word "co-imagineering" to rethink the notion of authorship.

María Mencía, author of the above-mentioned collaborative wreading Transient Self-Portrait, designed a module titled "Creative Digital Environments" that "engages students with a diversity of digital creativity in hybrid fields of electronic literature (e-lit), e-poetry and media arts". In this course, as María Mencía explains, students perform appropriative, as well as networked wreadings:

> pioneering works of media art and their historical influences; change and hybridity in electronic literature (e-lit); interweaving methods, content and technology in the production of creative practice; integration of theory and practice in practice-based research; culture remix and digital creations; creative programming and collaboration; concepts of appropriation, authorship, publishing and dissemination, and with this, the Creative Commons. (133-134)

The course taught by Mencía also "addresses media archaeology and methods to archive and preserve digital media works," which means that thanks to practice-based research, students are not only able to critically address works, but also understand the process of writing, publishing and archiving them. The course provides students with the opportunity to compare conventional academic research and practice-based research, address political and social issues, share personal experiences, as well as examine the processes of writing and reading print and digital texts. Mencía's students are also introduced to "remix creativity", or what Lawrence Lessig named "read write (RW) creativity", by contrasting "readwrite" (RW) and "read-only" (RO) cultures (Lessig apud Mencía 134). The concept of authorship is thus analysed in connection with the process of appropriation and collaborative writing. During the course, Mencía resorts to digital tools and platforms, as well as open-source software, so that students can play "the role of the author/reader/writer/artist/scholar" (134).

Torres', Zamora's and Mencía's teaching experiences indicate that the concept of authorship continues to be subjected to creative re-examination. Inside classrooms, this review process mirrors a concern to achieve a balance between practice-based learning processes promoted by digital media and conventional reading and writing activities. Charles Bernstein, in his "Wreading Workshops," conducted wreading experiments (or "creative wreadings") which entail the following activities:

> interactive and creative responses to assigned readings, including imitations, memorizations, rearrangements, word or phrase substitutions, homolinguistic (English to English) translations/ transpositions, and other "wreading" experiments (reading through writing) that involve reordering or rearranging a poem's words, lines, or stanzas, as well as locating or isolating certain key linguistic, figurative, and rhetorical features of the poem. (276)

Bernstein believes that the goal of these wreadings is "to investigate the recombinant structure of a poem—to what extent it retains its identity through modification of its constituent elements—and also to allow for a more intuitive, even visceral, contact with the materials of the poem" (ibid.). Therefore, to Bernstein, the process of reading a poem is not related to the activity of uncovering its hidden truths or finding an ultimate explanation, but with performance, intuition and, as stressed by Bernstein, disagreement. Reading a poem is thus a collective and interactive experience, rather than an individual and systematic process. According to Bernstein, "you can't interpret what you don't experience" (ibid.).

Works Cited

Aarseth, Espen. *Cybertext: Perspectives on Ergodic Literature.* Baltimore and London, The Johns Hopkins University Press, 1997.

Allen, Michael. "This Is Not a Hypertext, But... A Set of Lexias on Textuality." *CTheory,* 2003. ctheory.net/ctheory_wp/this-is-not-a-hypertext-but/. Accessed 26 May 2020.

Amato, Joe. "A Re/view of Bolter's Writing Space." *EJournal,* vol. 1, no. 2, 1991. www.textfiles.com/magazines/EJOURNAL/ej1-2.txt. Accessed 26 May 2020.

Amerika, Mark. "Blurring Practices: The Work of Art As Public Offering." *ALTX online network,* 2000. www.altx.com/amerika.online/amerika.online.5.4.html. Accessed 26 May 2020.

Andrews, Jim. "Nio & the Art of Interactive Audio for the Web." *Electronic Literature Collection,* vol. 1, 2006. collection.eliterature.org/1/works/andrews_nio/The_Art_of_Interactive_Audio.htm. Accessed 26 May 2020.

Bernstein, Charles. "Creative Wreading: A Primer." *Poetry & Pedagogy. The Challenge of The Contemporary,* edited by Joan Retallack and Juliana Spahr. New York, Palgrave MacMillan, 2006.

Campos, Augusto de, Décio Pignatari, Haroldo de Campos. *Teoria da Poesia Concreta. Textos críticos e manifestos (1950 -1960).* São Paulo, Livraria Duas Cidades, 1975.

Clifford, Allison. "The Sweet Old Etcetera." *Electronic Literature Collection,* vol. 2, 2006. collection.eliterature.org/2/works/clifford_the_sweet_old_etcetera.html. Accessed 26 May 2020.

Cornis-Pope, Marcel and Ann Woodlief. "The rereading/rewriting process: Theory and collaborative, on-line pedagogy." *Intertexts: Reading Pedagogy in College Writing Classrooms,* edited by Marguerite H Helmers. Lawrence Erlbaum, 2002, pp. 146-164.

Eco, Umberto. *The Open Work.* Trans. Anna Cancogni. Cambridge, Harvard University Press, 1989.

Emerson, Ralph Waldo. *The works of Ralph Waldo Emerson. Vol. 3, Essays, second series.* Boston, Houghton, 1909.

Funkhouser, Chris T. *Prehistoric Digital Poetry: An Archaeology of Forms, 1959–1995.* Tuscaloosa, The University of Alabama Press, 2007.

Genette, Gérard. *Palimpsests: Literature in the Second Degree.* Trans. Channa Newman and Claude Doubinsky. Lincoln, University of Nebraska Press, 1997.

Glazier, Loss Pequeño. *Digital Poetics: The Making of E-Poetries.* Tuscaloosa, The University of Alabama Press, 2002.

Goldsmith, Kenneth. "Introduction." *OBJECT 10, Cyberpoetics,* vol. 1, 2002, pp. i-ii, www.ubu.com/papers/object.html. Accessed 26 May 2020.

Hines, Thomas Jensen. *Collaborative Form: Studies in the Relations of the Arts.* Kent, Ohio and London, The Kent State University Press, 1991.

Idensen, Heiko. „Could Poetry be made by all? Hypertextual Discourses-Technics: annotation, crosscultural linking, collaborative textnetworks." *Word Circuits,* 1999. www.wordcircuits.com/htww/idensen1.htm. Accessed 26 May 2020.

Klaiber, Isabel. "Wreading Together the Double Plot of Collaborative Digital Fiction." *Analyzing Digital Fiction,* edited by Alice Bell, Astrid Ensslin and Hans Kristian Rustad. New York and London, Routledge, 2014, pp. 124-140.

Landow, George. *Hypertext: The Convergence of Contemporary Critical Theory and Technology.* Baltimore and London, The Johns Hopkins University Press, 1992.

Landow, George. *Hyper/Text/Theory.* Baltimore and London, The Johns Hopkins University Press, 1994.

Luzón, M. José and M. Noelia Ruiz-Madrid. "Learning to Learn in a Digital Context: Language Learning Webtasks for an Autonomising 'Wreading' Competence." *CORELL: Computer Resources for Language Learning,* no. 2, 2008, pp. 28-45.

Mencía, María. "Creative Process: Interweaving Methods, Content and Technology." *Digital Media and Textuality: from Creation to Archiving,* edited by Daniela Côrtes Maduro. Bielefeld, [transcript] Verlag, 2017, pp. 133-150.

Menezes, Philadelpho. "From Visual to Sound Poetry: the Technologizing of the Word." Face, vol. 1, 1998, web.archive.org/web/20000411044844/http://www.pucsp.br/~cos-puc/face/s1_1998/poesia2.htm. Accessed 26 May 2020.

Ong, Walter J. *Orality and Literacy: The Technologizing of the Word.* London, Methuen, 1982.

Portela, Manuel. "Autoauthor, Autotext, Autoreader: The Poem as Self-assembled Database." *Writing Technologies,* vol. 4, 2012. www.ntu.ac.uk/__data/assets/pdf_file/0030/827085/Manuel-Portela-Autoauthor,-Autotext,-Autoreader-The-Poem-as-Self-assembled-Databas.pdf. Accessed 26 May 2020.

Rasula, Jed. *This Compost: Ecological Imperatives in American Poetry.* Athens and London, The University of Georgia Press, 2002.

Rau, Anja. "Wreader's Digest - How To Appreciate Hyperfiction." *netzliteratur.net,* 2000. www.netzliteratur.net/rau/rau_wreader.htm. Accessed 26 May 2020.

Reis, Pedro. "Videopoesia: Produção Poética Híbrida em Língua Portuguesa." *Poesia Experimental Portuguesa: Contextos, Ensaios, Entrevistas, Metodologias,* edited by Rui Torres, Edições Universidade Fernando Pessoa, 2014, pp. 101-116. po-ex.net/taxonomia/transtextualidades/metatextualidades-alografas/pedro-reis-videopoesia/. Accessed 26 May 2020.

Rettberg, Scott. "All Together Now: Collective Knowledge, Collective. Narratives, and Architectures of Participation." 2011, retts.net/documents/cnarrative DAC.pdf. Accessed 26 May 2020.

Rheingold, Howard. *The Virtual Community: Finding Connection in a Computerized World.* Secker & Warburg, 1994.

Rosenberg, Martin. "Contingency, Liberation and the Seduction of Geometry: Hypertext as an Avant-Garde Medium." *Perforations* 1.3, *After the Book: Writing Literature/Writing Technology,* edited by Richard Gess. noel.pd.org/topos/perforations/perf3/geometry.html. Accessed 26 May 2020.

Ryan, Marie-Laure. *Narrative as Virtual Reality. Immersion and Interactivity in Literature and Electronic Media.* Baltimore and London, The Johns Hopkins University Press, 2001.

Shklovsky, Viktor. "Art as Technique," *Russian Formalist Criticism: Four Essays,* ed. Lee T. Lemon and Marion J. Reis, Lincoln, NE, U of Nebraska Press, 1965, pp. 3-24.

Simanowski, Roberto. "Death of the author? Death of the reader!." *Dichtung Digital,* no. 19, 2001. www.dichtung-digital.de/2001/09/30-Simanowski/index2.htm. Accessed 26 May 2020.

Suter, Beat and The Absolute Wreaders. "The Making of 'The Famous Sound of Absolute Wreaders'." *netzliteratur.net,* 2004. netzliteratur.net/suter/wreaders.html. Accessed 26 May 2020.

Todorov, Tzvetan. "Teorias da poesia," *O discurso da poesia.* Trans. Leocádia Reis and Carlos Reis, Coimbra, Almedina, 1982, 7-14. [Originally published in *Poétique – Revue de Théorie et d'Analyse Littéraires* 28, 1976].

Zamora, Mia. "Distilling the Elements of 'Networked Narratives' with Digital Alchemy." *Digital Media and Textuality: from Creation to Archiving,* edited by Daniela Côrtes Maduro. Bielefeld, [transcript] Verlag, 2017, pp. 151-165.

Chapter 2

Community Code: A thing, something, everything and nothing

Taeyoon Choi

School for Poetic Computation

Abstract: In *Community Code: A thing, something, everything, and nothing,* Taeyoon Choi depicts an abstraction of the natural world in an attempt to guide inquiries about how we can thrive in symbiosis with the magnificent living and non-living beings of the earth. In this exploratory essay he examines the poetics and politics of code, through artistic examples, metaphors and terrestrial inquiries. How can we create meaningful change in the Community Code, in order to transform our relationship to each other, technology, society and culture? Can computer codes become a tool for poetic and political exploration of being together? Can we reverse-engineer the tools that are traditionally used for oppression and exclusion of marginalized populations through gatekeeping and the hoarding of power, data, and resources? Can we unlearn the codes of control and manipulation to rethink all of our relationships to technology and the environment? Can we build social justice infrastructures, not limited to sociality or proximity, but in recognition of a global commons that we build and maintain for ourselves and others?

Keywords: Community, Code, Poetic Computation, Protocols, Technical Objects, Symbiotic, COVID-19, Digital Relations

***

Foreword by Iretiolu Akinrinade

A conventional definition of *community* consists of people with shared regional ties or those who share culture, language, interests and traditions. In this chapter, Taeyoon Choi offers a response to the theme of *Community Code* by reflecting on his pandemic propelled move from New York City to South

Korea. He imagines the future along with his community of artists, technologists, and activists' visions for realizing radically different futures of distributed, communal care.

Utilizing knowledge derived from the social disruption that COVID-19 prompted, with the non-conformist sensibility inherent to artists, activists, and critical theorists, this essay intends to prompt the imagination of a broader sense of *community* across boundaries of race and nation, as well as perceived difference and lack of personhood. In adopting such a world view, we can gather practices of accountability to embed care within our codes of conduct and operations.

The COVID-19 pandemic and lockdown introduced global social fragmentation of perceived communities. Simultaneously, technologies became essential for survival as people relied on radio, television, internet, and wireless services for critical and rapidly changing information pertaining to the virus. The workplace, along with more casual socialization spaces, faced new boundaries: whether at the threshold of one's apartment door, or the geographic borders of states and countries–spaces only recently declared impermeable to citizens of certain privilege.

The differential impact of a socially distant world on people with various structures of life was quickly rendered more visible. Along with the inability for blue-collar and service industry workers to "work from home," the shortage of groceries, face masks, and essential medicine in March 2020 in the United States exposed a failure in social welfare, public health care, and the global supply chain. Additionally, the inability to social distance in government detention facilities, illuminated the pinnacle practice of incarceration as discarding society's most vulnerable to conditions conducive to death or increased socio-economic exploitation. Fortunately, in the moments of dire need, local community-based mutual aid initiatives became vital practice as many found themselves unable to rely on public infrastructures or commercial services. Although the term "mutual-aid" was likely just reaching the mainstream, these local, decentralized infrastructures have long existed to fill the crevices of American and human rightlessness inherent in our global economic system.

Without disregarding tragedies that have transpired this year, I am soothed to see more conversations of communal care coming to fruition in light of COVID-19. I would also like to emphasize the viability of this de-structuring moment as a restructuring one. In the same way that apoptosis, or neural death, occurs to facilitate new development in the brain, which ultimately permits synaptic remodeling and long-lasting neural connectivity, the deconstructing of our social lives in 2020 can permit the emergence of responsible social obligation to a much broader community. At the introductory Seminar on Protocols hosted by The Vera List Center for Art and Design and The New School, Chancey Fleet, assistive technology coordinator, described the technological response to COVID-19 by acknowledging it rests on assumptions of "*reflexive digitality.*" What some consider to be an easy transition can be a transition of unease for disabled people, including low-income and elderly communities with low or no internet connection or outdated devices incompatible with the newest technology. Along with communicating the information that is seemingly important to people not living with a disability, meaning not dis-abled through relevant means– we must listen to a broader community to determine what information is being withheld by barriers within our technologies. Have the respective creators considered that they are withholding visuospatial cues on Zoom and Google Meet, crucial for location and speaker information to someone who perceives without vision? Do creators of empathetic VR technologies hesitate before investing their time creating a system that commodifies pain and subjectivity? Do "disruptive technologists" consider obsolescence to be contextual, knowing all products end up somewhere? From production to disposal and the multidimensional ways in which creations are consumed, there are individuals, communities, and ecosystems that ought to be protected by our community code.

The consideration of community code ethics brings me back to the wisdom of Ruth Wilson Gilmore and other adherents of the Black Radical Tradition. This tradition, "is, inexactly, movement away from partition and exclusion, – indeed, its inverse" (Gilmore 237). The recognition of our interdependence and connectivity between beings with diverse experiences is foundational to strive for more equitable futures. Some critical theorists posit that building an equitable future is a daunting if not impossible task, citing exploitation and varying levels of personhood as crucial to the functioning of society (as we know it).

Even still, radical imagination is what we have to realize alternative ways of living with one another. The creative technologist is someone with the liberty to both imagine and actualize. What creative technologists generate is something similar to language, structured before and by the communicator. Taeyoon's investigation of the life and death cycle of technological objects, paired with the stewardship responsibility of creators, has been iterated across disciplines in different forms. In Toni Morrison's use of the allegorical wise blind woman tested by her tormentors, the wise woman appropriately states, "I don't know whether the bird you are holding is dead or alive, but what I do know is that it is in your hands" (Morrison). Where Toni, as a writer, reads the bird as language, in its multiple manifestations and applications, Taeyoon posits technologists can read it as code: of conduct, of operations, and the coded meanings in systems of code itself. From here, we can begin to code a truly inclusive digital future.

1. The New Normal World

Since co-founding the School for Poetic Computation in New York in 2013, I've been studying computer hardware and code, specifically programming languages, with a community of artists, educators, and engineers. The very definition of *Community* and *Code* is changing due to the global pandemic of 2019 – 2020. Rapid readjustment to recent confines has many people wondering about the enduring updates to our new normal. It feels as though we are entering a New Normal World. Therefore, it's timely to redefine *Community Code*: what are the *codes* for technology in a post-COVID-19 world? How are hackers, engineers, technology educators, and activists changing the way they write, share and publish their codes? What are the new concepts of *community* in the post-COVID-19 world? Can a community of those *who have nearly nothing in common*, come together to build and maintain the codes of our New Normal World?

First, let's consider who and what lives in the New Normal World. My imagination of this New Normal World is inspired by Donna Haraway's *Staying with the Trouble: Making Kin in the Chthulucene*. There is the living: humans, animals, critters, microorganisms, and the breathing ecosystem. And there is the nonliving: things that exist but do not breathe. Technical objects, a broad term inspired by Gilbert Simondon's *On the Mode of Existence of Technical Objects*, including computers, smart devices, and biotechnologies, can be understood as both living and nonliving. Like us, technical objects have prescribed life and death cycles.

If you are open to imagining our New Normal World, not as a singular future, but multiple futures, time, and space created through many people *worlding* their worlds, I invite you to imagine the future *Community Code* with us. The concepts of care, accountability, and stewardship are key in working through these questions. I imagine travelling from the core of the earth to the world of celestial objects, beaming through levels of abstraction, from the layers of the Thing, Something, Everything and the Nothing. I seek to comprehend the continuity of community within these different spaces of abstraction.

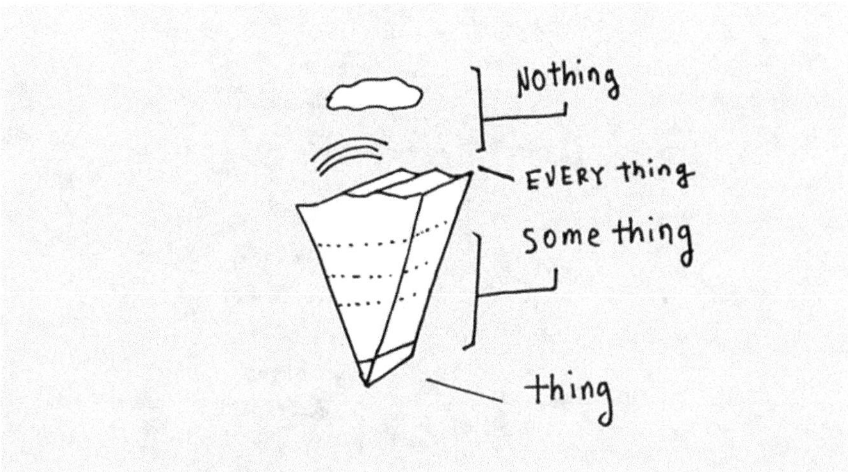

Figure 2.1. The Technical Object. Taeyoon Choi, Layers of computing #1, Digital illustration, 2019.

While it's common to think of computers as sleek metallic objects with a keyboard, mouse and screen, computers take on various material forms and perform a wide variety of operations. The "technical object," is not a material thing but a functioning: "The machine is a being that functions" (Simondon 151). Upon conceptualizing the technical object as a being with agency and a life form, we can come to an agreement that its components, its raw materials, have rights to be protected within the *Community Code*. Conversely, the technical object needs to be held to the values of *Community Code* when interacting with other members of our earthly community at large. Simondon's concept of technical objects is a relation of functions, "the technical being evolves through convergence and self-adaption: It unifies itself internally according to a principle of inner resonance" (ibid.). Through this inner resonance, movements of abstraction and repetition, the living and nonliving take up space in the earth and shape each other. The future *Community Code* may be a constellation of multiple living and nonliving beings, where the human connections with the natural resources and environment materialize as a distributed web of care. It will be a world which *worlds* a *caring world.*

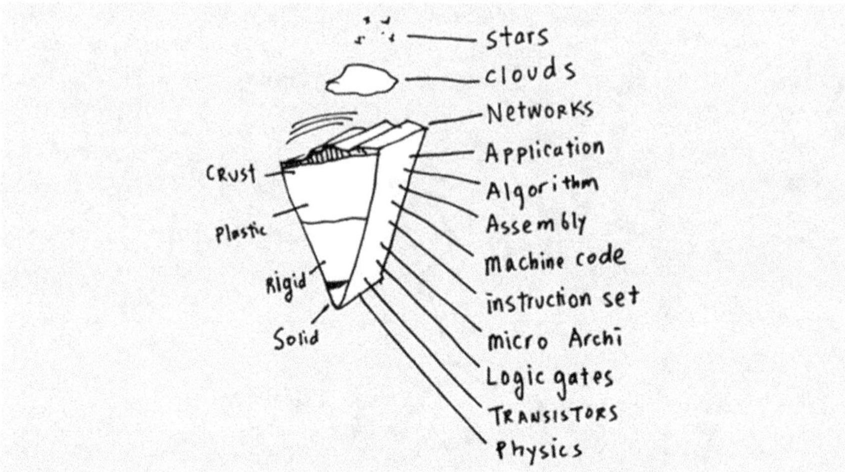

Figure 2.2. The Technical Object Abstracted. An illustration of technical objects in abstracted form. Electricity, for computers, is like the core of the Earth. It's existence and change of state is essential in making everything else exist. The binary state of on and off is like the existential state of being and non-being. In a similar way that the Zero and One are the foundation of most of the computers today built on Binary logic gates, computers scale to an abstraction machine based on Von Neuman architecture.
Taeyoon Choi, Layers of computing #2, Digital illustration, 2019.

2. Thing

Let us travel to a very hot and dense space to imagine a symbiosis with objects that are physical, technical, yet somewhat abstract. The inner core at the very center of the Earth is roughly 760 miles in radius and takes up a significant proportion of the earth's mass. Its temperature is around 9806 °F which is similar to the temperature of the sun's surface. The core's hot temperature and the gradual cooling of the Earth's interior generates magnetic waves (Alfe, Gillan & Price). Despite its significance in creating the Earth's magnetic field, the earth's inner core is still unknown to us. Much like the far away universe, galaxy MACS0647-JD, which is about 13.3 billion light-years away, a place no humans have ever visited, a place we only have a very limited information of, no human has ever been to the earth's core (Gutro, Weaver & Villard). Nonetheless, the earth's core is ever so present in our lives. If the core would not exist, the world as we know would disappear very quickly due to orbital decay, and we would be overexposed to the unbearable amount of radiation and solar wind (Lee). Unknown to us, we maintain a symbiotic relationship with the earth's core. From the core of the earth to the gravitational field it produces, our community consists of all beings that socially interact.

In Donna Haraway, *Staying with the Trouble: Making Kin in the Chthulucene*, sympoiesis applies to nature and culture that coexist in a system of co-creation. The living and nonliving, the biological and artificial, form interdependent relationships of "making-with." Recognition of sympoiesis requires a constant reconsideration of the life and death cycles of our human peers, and non-human helpers. We must reimagine our relationship to technical objects, the computers, the code that runs within as existing in symbiosis with the community who creates, maintains, and uses the technologies.

Sympoiesis is an enduring biological interaction between multiple biological organisms. In writing about the Wertheim sisters' crocheted coral reefs, Haraway speaks of coral reefs as a sympoietic system, a complex and organic entanglement of multispecies in coexistence. While computers are primarily nonliving electronic devices, we spend a tremendous amount of time interacting with them, as tools of communication, control, calculation, and entertainment; they impact our lives and hold characteristics and form that gradually adapts over time. After all, the computers come from the earth, its chips and hardware interfaces are made of rare earth minerals. Following their short life as consumer electronics or industrial machines, they go back to the earth, oftentimes as electronic waste that is difficult to recycle. Learning from Haraway's definition of sympoiesis, we can understand nonliving things that are always connecting to the living; we are co-creating a world with the nonliving, such as computers which use binary representations, in a flux between the existential state of being (One) and non-being (Zero). This flux is reflective of cycles of life and death, a circular progression of existence in the living and nonliving. To facilitate much of this interaction, most of the computers today run on electricity, the potential differences between the ground and the voltage.

Between the technical object and social thought lies conceptual and physical connectivity: social organization and creations determining that which is given life and between creations and use or abuse of resources that which is subject to death. The maintenance of this precarious eco-system of life and death is presently unmet in part due to conceptions of community and mutual responsibility that is limited to our immediate surroundings or human-kins.

adrienne maree brown wrote in her 2017 book, "Nothing in nature is disposable. Part of the resilience of nature is that nothing is wasted... there are toxic materials... and yet nothing is disposable, the cycle of life ultimately makes use of everything" (brown 131). Despite this truth, we collectively produce, consume, and dispose as if our concept of a product's death is real and resolute. About fifteen years ago, I had a studio next to an electronic waste facility in an industrial area near Ilsan, South Korea. Each day I saw trucks coming into the facility, full of consumer electronics such as TVs, desktop computers, household electronics and phones, that would contribute to an infinitely growing mountain of debris in the facility. Workers went through the pile, sorting between items that can be recycled or salvaged, and those deemed suitable for final discard via export to the Philippines and other parts of the global south. Learning how electronics are made from the rare earth minerals and the environmental impact of electronic toxic waste, I realized technical objects are also never separated from the earth and our bodies. We ought to appreciate our symbiotic relationship with technical objects through cycles of life and death, in theory and in practice.

3. Something

Although the computers we use today look different from computers from the last decade, the inner workings of the Operating System, especially the UNIX systems and commands have not changed very much since the 1960s, if not earlier. We often take UNIX for granted. However, some artists and scholars have raised questions about the cultural implications of UNIX. Tara McPherson connects the modularity in the UNIX systems to the social modularity essential to maintaining a system of racial segregation. McPherson writes, "This push toward modularity and the covert in digital computation also reflects other changes in the organization of social life in the United States by the 1960s. For instance, if the first half of the twentieth century laid bare its racial logics, from "Whites Only" signage to the brutalities of lynching, the second half increasingly hides its racial "kernel," burying it below a shell of neoliberal pluralism" (McPherson). Technical objects hold racial, gender, and other social biases of those who build them. If stewarded with care, technical objects can be used as tools of social cooperation and decentralized, grassroots organizing. For example, UNIX can be used to create non-oppressive Operating Systems, to forage tools of modularity towards emancipation, as opposed to oppression. I believe technical objects, even the ones that were originally created as tools of oppression, can be used to imagine and bring non-oppressive futures. There can be a way of life, a way of being with technical objects, that is different from our life with technological products today.

What would a nimbler network and node of the future technical objects look like? A collective called Black Quantum Futurism (BQF) declares their practice as "a new approach to living and experiencing reality by way of the manipulation of space-time in order to see into possible futures, and/or collapse space-time into a desired future in order to bring about that future's reality. This vision and practice derive its facets, tenets, and qualities from quantum physics and Black/African cultural traditions of consciousness, time, and space" (Black Quantum Futurism). Based out of Philadelphia, their work spans the intersection of music, digital media, performance, legal aids, and community support. BFQ's holistic approach to building community through ubiquitous Afro-futurist organizing helps me to shift in my thinking around technical objects towards a non-oppressive, social justice-oriented *Community Code.*

The COVID-19 pandemic was a moment of revelation of the severely precarious codes that run the society today; the legal codes, technical codes, and commercial codes of operations. During the momentary slowing down of the highly synchronized, techno-futuristic society, our attention centered around archaic forms of self-perseverance; washing hands properly, stitching scrap fabrics to make face masks, physically distancing ourselves from one another to limit the spread of the infectious virus. There's poetry in rediscovering low-tech solutions in a high-tech world, as the poetry speaks of the most emergent and necessary steps for collective survival. If the conventional relationship to technology were based on human use of the tools, consumption of the products, experience of the spectacles, we can think of different *Community Code* for the new world. It may begin with technical objects' entanglement with the community, superposition of living and nonliving entities, and permitting the oscillation of multiple times and spaces in coexistence. These interdependent, co-existence of technical objects, creating of habitats, and the shift between creative technology to community technology takes place in the Everything layer.

4. Everything

COVID-19 created a state of emergency for companies and states to rapidly adjust their legal and financial operations. The slowing down of capitalist machines during the quarantine and lockdown was a revelation of the various relationships we have, contribute to, and maintain. The code we use reflects and determines the society that exists today. When we think of protocols, we generally think of computer protocols, such as TCP/IP, HTTP, the grounding rocks of the internet and decentralized communications. It's true if you want to understand how the world is communicating these days, and I mean 'the world' metaphorically as the bespoke communication of machines and databases, you need to read_AWS's protocols or Zoom's_API documentation. These documents hold both technical and social *Community Code.* How can we change the *Community Code* to a future that's more equitable? I present three examples.

One. Protocols are a list of agreements and rules which we encode to machines, which are essentially all physical computers, servers consuming electricity somewhere. In *Computing, Climate Change, and All of our Relationships* Nabil Hassein argues for an entanglement as not limited to humans and biological lifeforms but yearning towards the interdependent relationship with the nonliving entities, such as technologies. Hassein prompts us to rethink the meaning of (technical) code and (digital) community in the context of decolonization. Decolonization is understood as the arduous work of slowly changing our thinking, our languages, our codes, and our communities' relationship to power. In technology this begins with an active de-centering the White and western understanding of ourselves, the humans as "masters of tools." It's important to think about the corporeal aspects of code, ecologies, and our relationships to living and nonliving entities. His call to action, to consider all of our relations as essentially interdependent webs, our complicities are very much a part of the ongoing crisis.

Two. The semantics of protocols matter, because they reflect the society that uses the protocols. In light of protests of police brutality and systemic racism in the United States, GitHub renamed their 'master' branch with the neutral term 'main' to abandon reference to the language of slavery (Cimpanu). On a practical level of linguistic decolonization, the name change can communicate care in code outside the very white, male programming archetype. On a physical level, semantic shifts can facilitate a reorientation of our bodies and minds with the earth and others. There are open debates about if corporate semantic shifts are efficient or performative and if the terms are even factually correlated with Black subjectivity. I think the real debates are to be had are between ourselves. What is it about mastery – power and domination – that is inconducive to collaboration, *how* does it, as Morrison put, 'limit knowledge' in this context? Mastery undermines the life in technology, conceptually limiting its life form. Just as we can take agency in language, programmers and technologists have a responsibility to take agency in rewriting coded language, and coding in alternative languages to better reflect the collaborative, multifaceted world we want to live in.

Three. Introducing beneficent protocols requires imagination and action. Artist and researcher Francis Tseng compiles a collection of resources on Aren.na channel *State of Emergency: Things We Allegedly Can't Have Except That We Can.* The site is an accumulation of resources that were believed to be scarce, unattainable, or high cost, which became available to those who were initially unable to access it. The compilation of media related to 'radical' practices in progress, such as abolishing the SAT and ACT or methadone distribution to homeless people in a pandemic, are the very means of informing what can move from speculation to reality. In the New Normal World, we can live in a world beyond artificial scarcity, and actualize a world of radical generosity, manifested as love in the Nothing layer.

5. Nothing

In the Spring of 2020, teachers at the School for Poetic Computation (SPFC) responded to COVID-19, and the movement for Black Lives and organizing against police brutality in the U.S. through their art, social justice organizing and community-oriented mutual-aid projects. They sought out new ways of connecting with strangers far away and discovering a profound sense of friendship with someone they never met in real life. The school offered a wide variety of online courses for students in various places, all joining remotely for the first time. "What if all the software we used was made by people who love us?" was a leading question posed by Melanine Hoff, an artist and teacher at SPFC, led a 10-week, online class Digital Love Languages with the class stewards Max Fowler, Adina Glickstein, and Amber Officer-Narvasa. The students of Digital Love Languages, mostly first-time coders, learned the UNIX folder structure, web programming and Natural Language Processing. They created unique software for their friends, esoteric websites, collaborative playlists and various forms of care. I understood their attempts to connect with like-minded collectives to overcome the sensation of isolation. Their work inspired me to imagine the "Internet that loves us." The conventional software we use, such as text editors, or social networking apps, are developed by people who don't know us, for a group of unidentified users. Instead of creating profit, what if our software was created to express love? The idea of love is commercialized and weaponized in today's world. Even in the apps that are designed to help us find "love," the experience is often transactional and explicative. In "Digital Love Languages," the idea and practice of love is not limited to romance. Scott M. Peck said, "Love is an act of will… both an intention and an action." In Community Code, love is creating a safe space for learning, to build a community of allies and organizers, and to share our thoughts via code through profound honesty. In the post-COVID-19 world, we can imagine a world beyond artificial scarcity, to actualize a world of radical generosity. COVID-19 and the temporary slowing down of the worlds, the protests and organized actions against the prison industrial complex, and the wide awakening of the systemic racism in our organizations, shows us that other worlds already exist; it's possible and necessary to change our worlds. That would be the world which worlds a caring world.

6. Epilogue

I wrote this essay between July and October of 2020, between New York City and rural South Korea, with assistance and in conversation with Iretiolu Akinrinade who was splitting time between New Orleans and Chicago. This essay is an attempt to expand the definition of community to inclusive of the living and nonliving: animals, critters, composts, humans, waste, and technical objects. Code is the software, hardware and code of conduct that shape our material realities. Community and code are never separate. Therefore, I wrote about the artists, designers, technologists, and philosophers in my community who've inspired me to rethink my relationship code. COVID-19 brought a chance to reset our Community Code, the modes of operation in our organizations, schools, museums, technologies, and our relationships with each other. A moment of emancipation begins with the code of love, care, and mutual support.

Due to the COVID-19 pandemic, I decided to relocate to South Korea, moving out of New York which has been a personal and professional home for the last decade. I am located in Seoul now, but I plan to eventually return to a mountainous village where I spent time in my childhood. I imagine farming, engaging with local communities and continuing my art and teaching practice there. While it's tempting to turn to a romantic vision of new rural life, I'm aware that "going back to where I came from" is not simply a reverse movement to a place of innocence. The reverse migration, a pattern that's common among city dwellers returning to their hometown or country of origin, also results in the world-as-we-know becoming less integrated. It's easy for us to hide in our cultural comfort zones, setting up visible and invisible walls around our national, cultural, linguistic, socioeconomic borders. For those who are searching for Community Code in the New Normal World, the challenge is to keep our creative and cultural practices open to new ideas, different identities, and work towards solidarity, while keeping our daily practices local, less consumptive, and less wasteful. How can we stay connected and work together, when we are far away and can't come together in a conventional sense? The world we know today, is one possible manifestation of many possibilities. It's time to imagine and live in multiple worlds. Technical objects can be used as tools of exploitation, or as vehicles of emancipation. If we believe in multiplicities, acceptance and inclusion of diverse identities and interdependencies, we can create technical objects that have symbiotic relationships with the living and nonliving, here-and-now.

How can we create meaningful change in the Community Code, in order to transform our relationship to each other, technology, society and culture? Can computer codes become a tool for poetic and political exploration of being together? Can we reverse-engineer the tools that are traditionally used for oppression and exclusion of marginalized populations through gatekeeping and the hoarding of power, data, and resources? Can we unlearn the codes of control and manipulation to rethink all of our relationships to technology and the environment? Can we build social justice infrastructures, not limited to sociality of proximity but recognition of a global commons that we build and maintain for ourselves and others? Ultimately, can we consider ourselves as the stewards of cultural commons, inhabitants of the New Normal World? May these queries guide your worldview, your life, and your practice.

For us, those who are seeking new ways of life after the traumatic events of 2020, Community Code is a call for action towards liberation of the people who are traditionally marginalized by the codes of society and a striving for beneficence towards our collective home on earth. For those of us who were not granted equitable community membership through systems of exclusion and alienation, by intentional neglect and tokenization, and those who empathize with all marginalized by race, class, disability, gender, and sexuality, here is our chance.

Works Cited

Alfè, D.; Gillan, M. J.; Price, G. D. (2007). "Temperature and composition of the Earth's core." *Contemporary Physics.* 48 (2): 63–80. doi:10.1080/0010751070 1529653.

Black Quantum Futurism. "About BQF." *Black Quantum Futurism.* www.black quantumfuturism.com/about. Accessed 1 Oct 2020.

brown, adrienne maree. *Emergent Strategy: Shaping Change, Changing Worlds,* AK Press, 2017.

Cimpanu, Catalin. "GitHub to replace "master" with alternative term to avoid slavery references." ZDNet US Edition. June 14, 2020. www.zdnet.com/ article/github-to-replace-master-with-alternative-term-to-avoid-slavery-re ferences.

Fleet, Chancey. Seminar 1: Protocols as Language and Communication. As for Protocols Seminar Series. 14 Sept. 2020, Vera List Center for Art and Politics Online.

Gilmore, Ruth Wilson. "Abolition Geography and the Problem of Innocence." *Futures of Black Radicalism* by Alex Lubin and Gaye Theresa Johnson, Verso, 2017, pp. 225-240.

Gutro, Rob, et al. "NASA Great Observatories Find Candidate for Most Distant Object in the Universe to Date." Edited by Robert Garner, *NASA,* NASA, 15 Nov. 2012, www.nasa.gov/mission_pages/hubble/science/distance-record.html.

Haraway, Donna Jeanne. *Staying with the Trouble: Making Kin in the Chthulucene.* Duke University Press, 2016.

Hassein, Nabil. "Computing Climate Change, and All Our Relationships." Talk delivered at Deconstruct 2018, Seattle, Washington 23-24 April 2020, Transcript available at www.zdnet.com/article/github-to-replace-master-with-alternative-term-to-avoid-slavery-references.

Hoff, Melanie. *Digital Love Languages,* School for Poetic Computation, 2020, www.lovelanguages.melaniehoff.com. Accessed 15 Sept 2020. The class zine can be accessed at digitallovelanguages.github.io.

Lee, Ezmeralda. "How Do Solar Winds Affect the Earth?" *Sciencing,* 2 Mar. 2019, sciencing.com/solar-winds-affect-earth-4566990.html.

McPherson, Tara. "Why Are the Digital Humanities So White? or Thinking the Histories of Race and Computation." *Debates in the Digital Humanities,* Ed. Matthew K. Gold. Minnesota: University of Minnesota Press, 2012, dhde bates.gc.cuny.edu/read/untitled-88c11800-9446-469b-a3be-3fdb36bfbd1e/section/20df8acd-9ab9-4f35-8a5d-e91aa5f4a0ea.

Morrison, Toni – Nobel Lecture. NobelPrize.org, Nobel Media AB 2020, Accessed 27 Sept 2020, www.nobelprize.org/prizes/literature/1993/morrison/lecture.

Peck, M. Scott. *The Road Less Traveled and Beyond: Spiritual Growth in an Age of Anxiety.* Touchstone, 1998.

Quartz Staff, "Tricia Wang thinks the future will be hyperlocal." *Quartz.* June 30, 2020. https://qz.com/1873417/tricia-wang-thinks-hyperlocal-collaboration-thrives-after-covid.

Simondon, Gilbert, et al. *On the Mode of Existence of Technical Objects.* Univocal Publishing. 2017.

Tseng, Francis. "State of Emergency." *Are.na,* 2020, www.are.na/francis-tseng/state-of-emergency-iua7pmsb2zc, Accessed 10 Oct 2020.

Wertheim, Margaret and Wertheim, Charlotte. "Art - Crochet Coral Reef." *Margaret Wertheim,* www.margaretwertheim.com/crochet-coral-reef.

Chapter 3

Cicada

Charles Baldwin

Independent Scholar

Abstract: Typically, the words collaboration and network inspire a sense of open communication among a plurality or collective in order to pool resources for a more effective and successful outcome. In this chapter, Baldwin explores the other side of online collaboration where networked interaction carries aims steeped in secrecy and cryptic engagement. The "net" has sponsored right from its beginnings activities better left hidden. In these collaborative worlds, agents adopt code names and work as much against each other, as with each other to solve mysterious riddles. The online world, we learn, still prospers from puzzle-solving games and narratives as part of its primary infrastructure. Baldwin offers a fascinating narrative on one of its most enduring interactive mysteries.

Keywords: Codes, Decoding, Puzzles, Online Cults, Gaming

***

What never was seen, or heard of, may yet be conceived; nor is any thing beyond the power of thought, except what implies an absolute contradiction. – David Hume, An Enquiry Concerning Human Understanding

Minister of Defense: You realize what you are implying? That we owe our human condition here to the intervention of insects.

Professor Bernard Quatermass: I suppose I am. – Quartermass and the Pit aka Five Million Miles to Earth (1968)

And so, starting at the beginning is difficult. Periodicity, punctuation: a case appears, a document to read, a community is at work, a network hums. But there is another case and another document and another community and another network. There are ripples and bubbles in all that is flat and continuously on. What is the starting point, when and where? Periodic cicadas remain hidden for years until they appear and swarm. Undoubtedly "the most anomalous and interesting of all insects" declared Charles Lester Marlett, who chronicled cicadas between 1898 and 1909 (7). He tracked their sudden appearance and loud vibrations, creating a classification of thirty "broods" – the word means breeding but also heat and vapor – hidden and then appearing across the eastern United States, each labeled with a Roman numeral from I to XXX, resulting in wonderfully death metal-like names such as Brood X. The classification is determined in large part by the cicada's periodic lifecycle: dormant for anywhere between thirteen and seventeen years and then emerging for four to six weeks to mate and lay eggs. Moreover, some of the thirty broods are as yet unobserved. They are hypothetical, they are yet to emerge or already emerged, they are among us, part of a periodic emergence, part of a latency of that which remains hidden and subterranean and yet active. The net is everywhere and nowhere, and the cicadas are singing.

"Their words had wings" said Homer (3.470). To tell a story, to philosophize, to boot up the net, to hear the cicadas sing. In the *Iliad*, Trojan elders are described as like cicadas on a tree branch. More specifically: their commentary is like the chirping of the cicada, and in turn, the cicada's chirps are famously word-like. In between significance and noise, in between politics and sound waves. The insect swarm is a boundary of becoming meaningful of the world and (inversely) the becoming noise of meaning. Cicadas have long fascinated, and not just for their sound. The extended cycles of underground hive life followed by swarming emergence and breeding led the insect to symbolize longevity and even immortality. Their swarms are images of networking, of hive mind. The red-eyed cicadas appear in Plato's *Phaedrus*: they sing continually without food or drink, a permanent background of song in the world. In doing so, they watch mankind and when the cicadas die "they go to the Muses and report who honors each of them on earth" (259c). What is this myth? Insectoid surveillance of our inspirational embodiment of the muses. Noise in the world and chaos that itself registers our actions and judges us. The world producing song, significance, information, not for us to read but to observe us, and we at best hear of the results in songs and in our fate.

There is no consensus among critics on what to make of the cicadas in Plato (Swanstrom 108-111). Cicadas are tied to the role of voice and of proper speech, the major concern of the *Phaedrus* dialogue, but not in any clear manner. Our speech and action are reported by cicadas but they give no rule or ethic for how to act. They regulate us, monitor us, accompany us, but without an FAQ or user's manual. Cicadas are speculations outside of philosophical dialectic, intrusions from the outside. We cannot judge them; they judge us: they produce the noise around us and report on our actions. Cicadas carry voice away from a single body into a swarm, into a cloud, into an elsewhere, and ultimately to the Muses, which is to say, into the medium. The piercing red eyes are just as important, as remarked in the texts. They watch and listen. The cicada's glowing eyes are sourceless, barely glimpsed, never the brilliance of insight and illumination. The noise, the red glow: do you not see it every night from your computer, from your modem, from the wires? Are these eyes not another name for the always-on, the ubiquitous, the net? Yet another name so many names, all the names (Pschera 2016). Cicadas are here, an elsewhere we carry with us. As ever, a name for the non-edge or inside-outsideness of information.

Starting over.

On January 5, 2012, a gif image was posted on 4chan/x, the paranormal board, an image of the following text:

> Hello. We are looking for highly intelligent individuals. To find them, we have devised a test.
>
> There is a message hidden in this image.
>
> Find it, and it will lead to the road to finding us. We look forward to meeting the few that will make it all the way through.
>
> Good luck.
>
> 3301

The text and the test, and all that followed, came to be called Cicada 3301 by the community of "solvers" that grew around it. Cicada 3301 continues to be given labels such as "the most elaborate and mysterious puzzle of the internet age" (Scott). Puzzle as in game but also as in unresolved enigma. It swarms across the net without answers and continues the cicada's complex song and periodic appearance. It appeared first in 2012 and repeated for years after, before falling silent, no new puzzles since 2014. The brood is waiting. Silence is an edge of the field where the cicadas are silent, latent, timing us: we are under observation.

What "is" Cicada 3301? A test or puzzle, a fiction, an artwork, an ARG, a government recruitment, a conspiracy, or a cult? Is it solvable or is it an activity and an experience that resists solution? The case and the documents are there to be read but the trails are unclear and disappear into controversy and contradiction. Anomalies produce myths. The lines of the cicada's song, its purpose and thesis, remain obscure and lost but the story can be told. Soon after the first 3301 message, regular posts surfaced on 4chan, Reddit, and other net forums attempting to "solve' the test. They, this loose group of "solvers," appeared and multiplied. They traded discoveries and deep dives. They buzzed with speculations and solutions. They collaborated. Along with requesting and sharing insights into individual puzzles, the solvers also shared theories about the origins and motivation of the test itself. The theories are a graph of possibility, of forms of labor, of appearance, and of readability. These theories are implicit in the language of puzzle, solution, test, and in the rhetoric of "we are looking for highly intelligent individuals." Many of the dedicated solvers got off the more public spaces of 4chan and Reddit for dedicated spaces such as IRC channels (the best known is #decipher). They lived and collaborated online. They were ever-present and never present. Most never knew the "true names" of their fellow solvers and kept to screen names such as Tekk. Solving Cicada 3301 filled their hours.

The fandom.com site "Uncovering Cicada Wiki" is the best starting point to reconstruct the story: it provides a detailed timeline and links to the evidence that continues to be available.[1] Maintained by the community of solvers, the "Uncovering Cicada Wiki" is itself part of the mystery. As a Fandom/Wikia site, "Uncovering Cicada" immediately situates Cicada 3301 in marketing space, surrounding the puzzles and clues with ads, but the anomaly remains. It starts with the cicada's song and the words on wings and fades but never ends in conflicting communities, laboring bodies, and click-through ad counters. Who are or were the community-solving Cicada 3301? Or rather: what does it mean to refer to a community, particularly around the anomaly of the cicada's song? To start but not finish these questions, here is the rough sequence of appearances and discoveries during the first Cicada 3301 puzzle, drawing in from various accounts, but primarily "Uncovering Cicada." Read it fast and make sure to stop for breath.

The image appears with a message to be read, including: "There is a message hidden in this image." "Message" here is not meant hermeneutically. The point is not that the image must be deciphered through reading but rather decoded technically. That is: the test immediately solicits skills in code and mathematics, though soon enough other skills will be required. It requires ways of seeing and solving that cross disciplines and histories. Technical and not hermeneutic: the distinction is difficult to sustain, the one breaks in

[1] https://uncovering-cicada.fandom.com/wiki/Uncovering_Cicada_Wiki.

another, there is never a return to a core of meaning and insight, of a subject that is informed and understands the song.

At the core of Cicada 3301 and other similar anomalous projects is advanced cryptography, frequently steganography (stego). Long used in practices of secrecy and coding, literally "hidden writing," stego is where data is hidden within other data. For example, the code of a JPG image contains many bits of data that are not displayed as part of the image and it is easy for an individual to add data, hiding a URL or a text or other message. To look for stego is to unsee the world as objects and forms, to let appearances dissolve into codes that may be hiding other codes. With the first 3301 image, all that is needed is to open the GIF in a text editor such as notepad and indeed you will find ASCII characters with appended, additional strings, including: "TIBERIVS CLAVDIVS CAESAR says 'lxxt>33m2mqkyv2gsq3q=w]O2ntk.'" The next step requires a more historical knowledge of codes: recognizing a Caesar cipher. Or perhaps not: as with much of the information required to "solve" the test, the answers are often available through Google searching and online tools. The key to solving and the membership in the community of solvers is less encyclopedic knowledge than ability to recognize clues, and then to search and manipulate data on the net, and then to keep moving, never to stop and read and dive in, always to look on for the next trace. In this case, for those solvers seeking clues, the name TIBERIVS CLAVDIVS CAESAR meant knowing enough to search and to arrive at the term "Caesar cipher." And what is this? More searching reveals that this substitutional shifting of letters for letters is one of the oldest ciphers, supposedly originated with Julius Caesar. Enough Google searching and soon the test solver arrives at a Caesar cipher tool that decrypts the code into a url: http://i.imgur.com/m9sYK.jpg. This turns out to be a link to another image, a wooden duck and text stating: "Woops just decoys here. Looks like you can't guess how to get the message out." For those determined not to end the test here, the task is to scrutinize the message for wordplay and to take the wordplay as a clue to a new trail, and so "get the message out" points to another stego software tool called Outguess. Using Outguess allows yet more hidden information to be extracted from the first image, as follows: "Here is a book code. To find the book, and more information, go to http://www.reddit.com/r/a2e7j6ic78h0j/ 1:20, 2:3, 3:5, 4:20, 5:5, 6:53, 7:1, 8:8, 9:2, 10:4, 11:8, 12:4, 13:13, 14:4, 15:8, 16:4, 17:5, 18:14, 19:7, 20:31, 21:12, 22:36, 23:2, 24:3, 25:5, 26:65, 27:5, 28:1, 29:2, 30:18, 31:32, 32:10, 33:3, 34:25, 35:10, 36:7, 37:20, 38:10, 39:32, 40:4, 41:40, 42:11, 43:9, 44:13, 45:6, 46:3, 47:5, 48:43, 49:17, 50:13, 51:4, 52:2, 53:18, 54:4, 55:6, 56:4, 57:24, 58:64, 59:5, 60:37, 61:60, 62:12, 63:6, 64:8, 65:5, 66:18, 67:45, 68:10, 69:2, 70:17, 71:9, 72:20, 73:2, 74:34, 75:13, 76:21." What is a book code? At the intersection of cultures of secrecy and literacy, of crypto and literary skills, it takes books as common objects in society, using a given book as a shared common key. The code is typically presented as clusters of three number coordinates, indicating a page number, a number of lines from the top of the page, and the number of words, all in order to locate the particular word or letter forming part of the enciphered message. Anyone with the actual book can find the letters or words indicated and reconstitute the code. So, the book code in the 3301 duck image is accompanied by a URL to a subreddit about the *Mabinogion*, a collection of Welsh stories and folklore dating from between the 11th and 13th centuries. Applying the book code to the *Mabinogion* leads to two images, each of which contains stego code revealed through the Outguess tool. Along with more clues, the messages are signed with a PGP key, ensuring secure and reliable communication with 3301. Using the clues to extract a string from the header of the subreddit and applying it as a "key" to the book code leads to the following message: "Call us at us tele phone numBer two one four three nine oh nine six oh eight." Calling the since deactivated phone number gives the following message: "Very good. You have done well. There are three prime numbers associated with the original final.jpg image. 3301 is one of them. You will have to find the other two. Multiply all three of these numbers together and add a .com to find the next step. Good luck. Goodbye."

Pause the transit here: many solvers were lost and fallen away by this point but some continued the labor of solving. Continuing on Reddit, 4chan, and backchannels, they shared solutions. Many possible prime numbers were located but the solution turned out to be the pixel dimensions of the original image (509x503). Here again was the use of another form of stego, based on physical dimensions rather than textual information. Multiplying the width and height by 3301 provided a URL. The website corresponding to this URL contained an image of a cicada and a fourteen-day countdown. This was the first appearance or emergence of the cicada associated with 3301: getting this far let the solver hear the song and recognize the brood's appearance. All the previous clues or steps of the test were "found," strewn about the net, as it were. In short, they were encountered as a static object or form but here, for the first time, the periodical cycle of the brood and of the solver came into sync. Cicada 3301 was alive and the time of solution was the time of the cicadas. The countdown and the stego message found in the image by applying Outguess led the solvers to wait until "17:00 on Monday, 9 January 2012 UTC." When the count reached zero on that day and time, the page was immediately updated with map coordinates, pointing to fourteen locations in five countries in four continents. For the first time collaboration was unavoidable, requiring the ability to access all the locations. Solvers could not work on their own. Coordination and collaboration led travelers to the locations, where they found posters on walls and lampposts with a cicada symbol and a QR code. Once discovered, the QR codes were shared by the solvers. Together, the codes led to a URL with yet another image. More cryptographic analysis of the image discovered more book codes, as well as a warning acknowledging the community and the many puzzle solvers: "You've shared too much to this point. We want the best, not the followers. Thus, the first few there will receive the prize." The book code eventually led to William Gibson's *Agrippa*, and solving this using Gibson's book led to a Tor website on the dark web. Messages on the site congratulated solvers on getting that far and asked them to set up a new, anonymous email address to communicate with Cicada 3301. Those that did so and sent email to Cicada 3301's PGP secure address received more messages, more steps, more stego in text, and audio, and images. The community of solvers continued to share clues but the song faded. Some solvers arrived too late and the website shut down. Solvers were warned not to collaborate or share details. Soon there was no more shared information and no known solution. The puzzle ended here or at least those who continued did not publicly tell what happened and the accounts that surfaced later are contested and apocryphal. It is hard to know the ending of the story. It is hard to know how to finish this account. "There are many false messages out there," declared Cicada 3301. A final message declared "We have now found the individuals we sought" and contained a public key for authenticating future messages as in fact from Cicada. The end.

The brood was underground. The community waited. The song was sung, until the next time. But then, in 2013, another message on 4chan and elsewhere. More stego, more math problems, more phone messages, more book codes, a new list of coordinates (eight locations in four countries). New variations appeared, with Twitter now incorporated into the puzzle. Once again, the trail ended. There seemed to be no official message. Again, in January 2014, a message appeared on 4Chan and other web forums, with an ASCII image of a cicada, and stating that the image contained a hidden message. "There is a message hidden in this image." The path was similar: through images, codes, books; but there was a major difference: the *Liber Primus* book was released, written in an alphabet of runes that had been part of the 2013 puzzle. The book contains clues and codes of all sorts, as well as what seemed to be Cicada's "philosophy." It has yet to be fully translated. *Liber Primus* appears to be the "answer" and Cicada's communications seemed to confirm this. The answer is always in a book but what answer and what book and answer to what? The end of the puzzle is the book. Or: puzzle and book are the same. Or there is no answer since there is no clear publicly available translation of the book. Further puzzles followed in 2016, with no reason given for the lack of a puzzle in 2015.

Cicada 3301 buzzed elsewhere, gaining a pop fame. Mark Danielewski, in *The Familiar*, Volume 1, references "bread crumbs a la Cicada 3301" (647). In 2018, IMDB announced a Cicada 3301 movie, still unreleased. The movie's tagline is “We are looking for highly intelligent individuals. Good Luck.” A coffee table version of *Liber Primus* is available on Amazon.

It is difficult to describe what happened next. According to some accounts, the successful puzzle solvers were taken to a forum on the dark web where they communicated with members of the Cicada group. They were told that the goal of the entire puzzle and recruitment was to find a team dedicated to creating software for privacy and freely shared information, particularly intended to protect whistleblowers. The software came to be termed Cicada Anonymous Key Escrow System, or CAKES. These are legends from former solvers: they joined, signing on as distributed, working-from-home, software developers. And so, the end of cicada the myth and the reward for the effort of solving is an intake by HR into an IT project of ongoing unpaid labor devoted to the open and anarchic goals of the community. Over time, so the story goes, the recruits grew bored and left the group. Eventually the site went dark. The CAKES project came to nothing. A winner of the second puzzle stated that he was asked to wait to be contacted but it never happened. The various Cicada 3301 websites were also taken down. The five solvers of the second puzzle claimed to confirm this. Or what happened was entirely different. “I was part of what you call 3301/Cicada for more than a decade,” an anonymous author wrote, describing a more sinister goal of destabilizing world governments, “and I’m here to warn you: Stay away” (“The Infamous Warning”). In this description, Cicada 3301 is an anarchist cult. Or none of this was the case. The stories are disinformation or fables. It is all unclear, hard to know where to go. As with every myth, the hero may return to the town and work in a shop or continue wandering through the wilderness following the hunt. Myth becomes a book and a glow cast from the darkness. In 2017, clues appeared indicating that a puzzle was underway but nothing explicit was posted on the forums and from there, the cicada’s song went silent again.

Solving Cicada 3301 puzzles requires a deep understanding of contemporary cryptographic technology, NSA/GCHQ-like or Turing-esque skills. It also involves hackerish devotion to web tools and forums. It also requires a comfort level with literature and language. And, importantly, it requires cooperation: at least the publicly available accounts show that even the most crypto-savvy puzzle solver cannot get down without some collaborators to provide insight and assistance. Your task: solve the puzzles, read the texts, and what do you find? A library and form of labor. The library is the denoted content of the Cicada tests: it is both bookish and hackerish. It is suffused with literary culture and online culture, and exists at the nexus of the two. The many literary source texts are a leisurely wander through an undergraduate syllabus on myth, imagination, and liberty. Along with the *Mabinogion* and Gibson’s *Agrippa,* Cicada tests include texts such as William Blake’s poems

"Song of Liberty" and "The Marriage of Heaven and Hell," as well as Blake's images of Nebuchadnezzar, Newton, and "The Ancient of Days"; Ralph Waldo Emerson's "Self-Reliance"; Aleister Crowley's *The Book of the Law*; and Douglas Hofstadter's *Godel, Escher, Bach.* Additionally, the tests include Anglo-Saxon Runes, Mayan numbers, Zen koans, music by Johann Sebastian Bach, and paintings by John William Waterhouse. Impressive enough of a list but the tests do not require interpreting or discussing these sources, nor is listening or viewing required beyond what is necessary to find the opening of the next clue. No, what is required is moving through and transforming. Solving is traversing. A solver needs to know or research enough to be aware of the existence of Blake's work, for example, but not to read or have read it. Or rather: a certain kind of reading is involved, where the solver scans for text strings, skims the text, looking for where it links to the next clue in the test. In short, all the content of the texts, with its mysticism and freedom, its natural origins and individual history, its contemporary buzz and its classical gravity, is not necessary to the solver.

Marcus Wanner, one of the solvers of the original 2012 puzzle, was "initially attracted to 3301 because they built the puzzle the same way as I or most other free software developers would have" (Turkel). The reading practice necessary for these tests is less that of academic hermeneutics and closer to the culture of hackers, otaku, LEET programmers, and the like. Indeed, these groups and their practices are all invoked in the test's reference to and use of 4chan, TOR browsers, PGP encryption, and similar tools. And to be clear: the community around these tools may be philosophically committed to the ideas in Blake and Emerson and the wider reading list of the Cicada 3301 tests. Blake's "If the doors of perception were cleansed everything would appear to man as it is, Infinite" finds good company on 4chan and elsewhere. Nevertheless, solvers of the Cicada 3301 tests are not engaged through ideas but through decoding and recoding. The texts matter for the labor, for the work of solving. The mythic texts are the mythic tasks, the works of Hercules: these are the acts required to traverse the material and be recognized as a "highly intelligent individual" who made it all the way through.

The mythic labor of information: consider the Daily Telegraph puzzles of 1942 (Chivers). After a series of letters to the Daily Telegraph claimed the crossword puzzles were too easy, the paper offered a prize to anyone who could solve the January 13, 1942 crossword in under ten minutes. It was exceptionally hard but there were five winners. Behind this publicity stunt was a strategy to recruit cryptographers for the military: the five received confidential letters inviting them to join the unit that eventually worked in Bletchley Park breaking the German Enigma codes under the direction of Alan Turing. The kind of thinking and problem solving involved in crosswords translated to the creation of the digital computer. Secret codes and secret groups go together in the forming of cyberculture. The technique has been used since. In 2004 Google rented billboards in Silicon Valley and posted a complex math problem. The solution led to a URL for a recruitment website and a job at the search engine giant. In 2013, the Government Communication Headquarters (GCHQ), the British top crypto agency similar to the USA's NSA, released a set of challenges intended as a recruitment tool, under the title "Can you find it?" In 2014, the US Navy's Project Architeuthis (Greek for giant squid) released Operation Sleeper Shark, a video game that was in fact a cryptographic puzzle. Solving the puzzle led to a chance at recruitment to be a Navy codebreaker. All the puzzles involved complex computation, pattern recognition, mapping of codes onto URLs, and similar activities. Project Architeuthis claimed to be inspired by Cicada 3301. *Rolling Stone's* account quotes the puzzle designers: "We knew about Cicada and were inspired by it [...] We know that's where our audience lives [...] They live online" (Kushner). The mythic labor is dedicated to living elsewhere, to infinite recoding, to a body of information. Not the content but the form of the tests are continual translation and re-coding, movement and flow. Traces of the imaginary of pure information, the matter of the net, the timeless cicadian song heard by the solvers that is there from before the net was commercial or government or social media.

It is difficult to know where to start but there are examples, cases to describe, from Webdriver Torso, to Satashi Nakamoto, to the 11B-X-1371 video. Compare another anomalous appearance: r/A858DE45F56D9BC9 or A858. It is first of all a visual phenomenon, an appearance in the space of Reddit. Numbers and letters. There are spaces as well. A string is typically 32 characters long, with the numbers from 0 to 9 and the letters never more than A through F. These strings are divided by spaces. In short, it appears to be a 64-bit hexadecimal code. It takes over the Reddit post. From 2011 until 2016, almost daily, from multiple users and multiple threads on Reddit, post after post, without explanation. The posts are similar, clearly related, but remain enigmatic. Another subreddit devoted to "solving A858" was full of dedicated computer scientists and cryptographers who offered decryptions including an image of Stonehenge, a thank you, and an excerpt of Max Lerner's *Actions and Passions: Notes on the Multiple Revolutions of our Times.* Participants in the "solving" thread understood the posts as social security numbers to track people, map GPS coordinates, or ASCII code. The final post in 2016 was a prime number: 12764787846358441471, in hexadecimal B125A2EA5F6B1DFF. Solvers sought the number everywhere, focusing on websites with these numbers. There was no "solution." In August 2015, an account suddenly posted to the solving thread, offering to answer questions about A858. It was A858 "itself" finally speaking. The answers were cryptic and added to the mystery. In March 2016, the following post appeared: "The A858 Project Has Concluded. You may unsubscribe." What followed was nothing, silence.

“I only hope A858 actually has a purpose,” said redditor jdaher. “I don’t particularly care if it is a good one” (Kushner). AMA, another redditor writes: “When the purpose is discovered or disclosed it will be obvious to all.” Augenwhimmel wrote: “It’s an experiment. There are several components (to put it at its crudest): the mathematical/decrypting part and a social factor. I still am sure that the A858 is not only looking for very intelligent people, but is also studying behaviour.” Swebonny insists that A858 is “a piece of art that uses the ‘digital world’ as its canvas.” Again, the form of the artwork and the labor of solving. Surely this is the cicada’s song? Calling to those who live online, the community with nothing in common but the net, or rather nothing and everything.

History points to the Markovian Parallax Denigrate, an even earlier anomalous appearance in sync with the first moments of the net. Here is a typically posting, from August 1996 and still to be found in the Google Groups archive:

> jitterbugging McKinley Abe break Newtonian inferring caw update Cohen air collaborate rue sportswriting rococo invocate tousle shadflower Debby Stirling pathogenesis escritoire adventitious novo ITT most chairperson Dwight Hertzog different pinpoint dunk McKinley pendant firelight Uranus episodic medicine ditty craggy flogging variac brotherhood Webb impromptu file countenance inheritance cohesion refrigerate morphine napkin inland Janeiro nameable yearbook hark

Of course, the Markovian Parallax Denigrate is exactly not A858. The former is words and the latter numbers and letters. There is no similarity. Or is there? At that point where numbers and words meet in the notion of sequence, the instability of counting or ticking and language, the narrative mess of Richard's Paradox or the Cretan Liar. Word and number, sign and index. The labor of making and unmaking. The word in the Markovian Parallax Denigrate is but a number in an algorithmic sequence. The numerical string in A858 is a word in an utterance yet to be spoken. What is the Markovian Parallax Denigrate?

We are talking 1996 around when Amazon and eBay went online and Yahoo was dominant, only three years after the first web browser and two years before Google is in business. There were about 100,000 web sites. The Daily Dot website calls the "Markovian Parallax Denigrate" "the Internet's oldest and weirdest mystery" (Morris). The name is taken from the title of each of a series of posts to Usenet groups in August 1996. Remember, Usenet was the pre-web bulletin board-like system dating back to the early 1980s. At this point, a single post still circulates but the basic idea is there: a spew of words with no explanation or context. Hundreds of these posts appeared and just as quickly it was over. The event was largely forgotten and then revived as the notion of "weird internet mysteries" began to develop around 2006 or so. This kind of retrospective view a decade later marked a certain maturity of the web but it also meant, at that distance, that little or nothing of the incident remained. The Markovian Parallax Denigrate is a myth of anomalous information, unfolding text or numbers as the source of the net. With its afterlife as one of the internet's oldest and weirdest mysteries, it becomes a kind of origin point for spam or possibly for experimental literary practice on the web. One account tries to explain the origin of these texts in an erratic CIA operative named Susan Lindauer but even this information, if true, provides no solution to the texts and their source or meaning: it gives a name to their authorship but no motivation. Certainly, Markovian invokes Markov chains, a mathematical model describing a sequence according to statistical probabilities, and which can be used to computer-generate readable texts. The suggestion may be true: the spew of words generated by a Markov chain, although there is little structure to the example, suggesting a very random model. Yanking the Markov Chain.

"Had the concept of software been available to me, I imagine I would have felt as though I were installing something that exponentially increased what one day would be called bandwidth, though bandwidth of what, exactly, I remain unable to say" (108). This is William Gibson writing of the impact of first reading Jorge Luis Borges' story "Tlon Uqbar Orbius Tertius." The concepts now available, bandwidth but perhaps nodes as well as subnets and Cicada 3301 and all the rest, are ways of saying what the story "opens." Gibson continues: "This sublime and cosmically comic fable of utterly pure information (i.e., the utterly fictive) gradually and relentlessly infiltrating and ultimately consuming the quotidian, opened something within me which has never yet closed." So, the opening: the case of Cicada 3301 is another way of being unable to say what open is never closed. It is hard to know where to start. Borges's story is a usual suspect as a precursor of contemporary cyberculture – for example, often invoked in discussions of alternative reality games – yet the unsettling aspect of the story in its telling – or rather in the opening that Gibson describes – is that it is not and does not describe a game but rather produces or emits the narrative of such a game. So too, Cicada 3301 offers the carapace of a game. Worth noticing is that Cicada 3301 arrived in the wake of the public interest in ARGs, with high profile instances such as the viral marketing of *The Dark Knight* or of Nine Inch Nail's *Year Zero*. Scholars such as Jane McGonigal were busy situating ARG's as world-changing: "reality is broken" and we need to make it more like a game. The premise of ARGs and similar world games, from *Ong's Hat* to the Jejune Institute, is that "this is not a game." The circularity of this game is "this is not a game" is the game, and so on, is once again the game. The delight of an ARG, or for that matter a digital poem or Turing-testable chatbot, is the confirmation of the symbolic world that proffers significance to programmed machines such as humans and computers. What appears by chance should be within certain parameters: surprise is expectation. Symbolic relations and structures of capital are in place before the first word. There is a graph-theoretical edge between ARGs and games but there is no edge connecting Cicada 3301 and any artwork or games. The play of obscuring and revelation is central to the complexity of ARGs but Cicada 3301 offers nothing of this. No curtain to raise as reality yields or caves in.

Just as it is difficult to know where the case starts, so it is difficult to know where to end. When do you end reading and writing? Is the lack of connection the latest connection? The term "myth" qualifies the troubled causality, the strange topology, and the labor given to feed the net. What is the point of all this myth if the outcome is just another trick of capital, of labor captured into another idiotic software project? To give all to nothing, to the search, to the hunt, to what is over there. The prey is beyond. On the one hand, a sad and pointed story of neoliberalism where the pleasure is in the effort. In this story: Cicada 3301 is a template for discovering, enticing, and snaring the most advanced laborers on the internet. The solving organizes "creative" practice. Overcoding and diving in, following the traces on the decoding path until there is no until. This is the job. On the other hand, how long will the story last and when did it start? In the process, in the decoding, in the unfolding of the information, in this practice, is not manipulation palpitation and stimulation? Stimulation of what? Does this manipulated, decoded, and intensified information set the body abuzz? Surely this is the cicada's song? The solver's hunt feeds the net, labor is given to production, to the Muses and gods.

Works Cited

Chivers, Tom. "Could you have been a codebreaker at Bletchley Park?" *The Telegraph*. October 10, 2014. https://www.telegraph.co.uk/history/world-war-two/11151478/Could-you-have-been-a-codebreaker-at-Bletchley-Park.html (accessed July 7, 2020). Unpaginated.

Danielewski, Mark Z. *The Familiar: Volume 1, One Rainy Day in March*. New York: Pantheon, 2015.

Gibson, William. *Distrust that Particular Flavor*. New York: Berkeley, 2012.

Homer. *Illiad*. Trans. By Ian Johnston. Vancouver Island University: 2010. http://johnstoniatexts.x10host.com/homer/iliad_title.html (accessed July 7, 2020).

Hume, David. *An Enquiry Concerning Human Understanding*. Cambridge: Cambridge UP, 2007.

Kushner, David. "Cicada: Solving the Web's Deepest Mystery." *Rolling Stone.* January 15, 2015. https://www.rollingstone.com/culture/culture-news/cicada-solving-the-webs-deepest-mystery-84394/ (accessed July 7, 2020). Unpaginated.

Marlatt, C.L. (1898). *The Periodical Cicada: An Account of Cicada Septendecim, Its Natural Enemies and the Means of Preventing its Injury, Together With A Summary of the Distribution of the Different Broods* (Bulletin No. 14 – New Series, U.S. Department of Agriculture, Division of Entomology). Washington, D.C.: United States Government Printing Office. OCLC 10684275.

Morris, Kevin. "The Markovian Parallax Denigrate: Unraveling the Internet's oldest and weirdest mystery." *The Daily Dot.* May 3, 2020. https://www.dailydot.com/society/markovian-parallax-denigrate-spam-mystery/ (accessed July 7, 2020). Unpaginated.

Plato. *Phaedrus.* (Trans. Harold N. Fowler. Cambridge, MA: Harvard University Press, 1925.) Perseus Digital Library. http://www.perseus.tufts.edu/hopper/text?doc=Perseus:text:1999.01.0174 (accessed July 7, 2020).

Pschera, Alexander. *The Animal Internet: Nature and the Digital Revolution.* Trans. Elizabeth Lauffer. New York: New Vessel Press, 2016.

Quartermass and the Pit. Directed by Roy Ward Baker, Hammer Film Productions, 1967.

Scott, Sam. "Cicada 3301: The Most Elaborate and Mysterious Puzzle of the Internet Age." *Metro.* December 16, 2013. https://metro.co.uk/2013/12/16/cicada-3301-the-most-elaborate-and-mysterious-puzzle-of-the-internet-age-4229892/ (accessed July 7, 2020). Unpaginated.

Swanstrom, Elizabeth. *Animal, Vegetable, Digital: Experiments in New Media Aesthetics and Environmental Poetics.* Tuscaloosa: University of Alabama Press, 2016.

"The Infamous Warning." *Uncovering Cicada.* https://uncovering-cicada.fandom.com/wiki/The_Infamous_Warning (accessed July 7, 2020). Unpaginated.

Turkel, Dan. "After over a year of silence, a mysterious online puzzle has returned with a cryptic tweet." *Business Insider.* January 13, 2016. https://www.businessinsider.com/cicada-3301-returns-after-year-of-silence-with-cryptic-tweet-2016-1 (accessed July 7, 2020). Unpaginated.

Part II.
From the Hive to the Classroom

Chapter 4

Digital Swarm Techniques: A Case Study to Teach Digital Collaboration and Disrupt Power Structures in Education

Maria Aladren

MCVTS Theatre High School

Abstract: The author proposes a "swarm pedagogy" as a critical approach to digital educational spaces, seeking to increase students' creativity and diminish fundamental digital inequalities. A more equitable and approachable digital pedagogy will increase creativity and use student/teacher collaboration to disrupt power structures in the classroom.

Starting in 2015, the author engaged in a series of pedagogical digital experiments to increase creative output and free creative impulses in students from resource-deprived demographics. Those exercises resulted in increased testing scores, student responses of greater self-confidence, and a large bump in college acceptances.

The exercises coalesced around a collectively agreed creative goal, a closed but anonymous digital space, an exploration/exploitation split in the process, a fluid, often anonymously created, hierarchal structure and organization, and a set of random attraction/repulsion rules.

The exercises' success points towards the possibility of creating a set of collaborative, creative techniques that use our digital structures for the benefit of students and teachers in our education system.

Keywords: Pedagogy, Collaboration, STEM, Improvisation, Performance

***

> A technology always has a history, and it has a politics. A technology likely has a pedagogical bent as well.
>
> –Morris and Stommel,
> *An Urgency of Teachers: The Work of Critical Digital Pedagogy*

In a conversation with Petar Jandric, Henry Giroux highlighted the contextual nature of our human thought when he said, “like any social theory, like any ideology, like any worldview, critical pedagogy needs to adjust to circumstances in which it finds itself” (Jandric, *Learning in the Age of Digital Reason* 152). It is a bit of a truism to state that our circumstances, as humans--and as communities–have changed since Paulo Freire began to apply critical theory to pedagogy. Those changes in our circumstances became more apparent over my roughly twenty-five years of teaching performing arts. During that time, I gradually grew frustrated at a barely grasped intuition that many of my students were experiencing life—and therefore learning—in ways radically different from the generation before.

In an attempt to breach this now gaping gulf, I began to experiment—under the shield of a chronic lack of curricular oversight for the arts and with the help of a supportive administration—with a series of exercises, artifacts, and approaches that took advantage of an incipient digital space I thought I understood.

The illusion did not last long. As I explored this rapidly changing digital landscape, it demanded that I alter my thinking in ways similar to the culture-shock I experienced as an immigrant to the USA.

1. A journey towards a Swarm Pedagogy: Creativity, Spaces, Collaboration, Anonymity

I started my secondary teaching career at a vocational school—a system that carried with it connotations of a lower class and lesser ability students. In my experience as a teacher and teaching artist, those children are spoken of in the same tones as in one of Jean Anyon's studies, "Just do your best. If they learn to add and subtract *[or speak clearly in public or remain quiet in a theatre (my words)]*, that's a bonus. If not, don't worry about it" (Anyon 7). In arts education, the concept of "untalented" replaces Anyon's concept of "lazy." No matter. The results are the same—children of lower-classes remain mainly the consumers of mass-produced art and entertainment; they are not seen as future art producers; instead, they receive the fully formed narratives of their own experiences from an elite group of artists who control said narrative. There is, thus, a "reproduction of the division of labor at work and in society between those who plan and manage [...] and the increasing percentage of the workforce whose jobs entail primarily carrying out the policies, plans and regulations of others" (Anyon 32). Creativity and the fostering of such type of thinking is in many instances seen as an auxiliary section or instruction or a "fun time" to calm the student body.

My classroom demographic during the last ten years of my tenure (N=230) skewed toward a mostly female-identified (64%, male-identified = 32%, other = 4%), racially mixed (White=50%, Latinx=24%, Black=23%, Asian=3%), and blue-collar (over 62% of these students would be first-generation college-goers). In the early years, I discovered that my students did not feel they had an artistic voice, even though they wanted to be artists—many of them had never attended any theatre plays, museums, dance events, or other cultural institutions. On the other hand, they readily created digital material and placed it online, even though they had a hard time considering it "art."

Searching for ways to open doors for their creativity, I began to experiment with simple and accessible digital spaces, spaces that students could easily reach with phones, iPads, and minimal computing power. I sought ways to use those spaces to disrupt power structures in the classroom and give my students a chance at an artistic voice. These spaces were familiar to students and offered enough distance from analog curricular structures to try new approaches. The uses were far from innovative (Wikispaces, Google Docs, and such have had their place in the classroom for a while). However, those simple, accessible, and contained digital communities allowed for a version of "collective collaboration"—the type of devised creation that South American theatre practitioners like Enrique Buenaventura, Augusto Boal, Osvaldo Dragun, and Griselda Gambaro engaged in before it migrated to the northern hemisphere.

In combination, those elements disrupted students' ideas of what kinds of stories they were allowed to tell, how to tell them, and what kind of knowledge they were allowed to find, construct, and research to tell them. It broke down some limitations, self and socially imposed, on their creative process. Used that way, digital technologies and digital spaces offered a way to disrupt power structures in the classroom and to allow students to try new forms and explore subjects, methods of composition, and language that they would not attempt otherwise. Those digital spaces allowed us to have a collective creativity that enriched students' sense of accomplishment and benefited their thinking. With their chance at anonymity, the collective digital spaces allowed students to become familiar with the sense of failure that accompanies most creative processes without succumbing to the feelings of "being a failure" that are rampant among our most vulnerable student population.

In 2015, I began collaborating with the New Jersey Institute of Technology and the McCarter Theatre to explore pedagogical intersections between the arts and STEM fields. As part of this experimentation, we brought art students to a panel of STEM professors, sparking in the process several writing projects and strengthening our idea that technology could inspire young artists as much as art could aid STEM students. One panelist, NJIT professor Simon Garnier[1], and his thoughts on swarm theory, particularly how simple organisms achieve complicated ends by self-organizing, became the seed to a new digital collaboration approach in the creative classroom.

The germinal idea of an "organization without an organizer" (Garnier, Gautrais and Theraulaz 4) struck at the core of classroom power dynamics. It began to translate into a series of exercises aimed to decentralize the creative process, take away personal investment in story, and crowdsource material before an editorial process--what the students called "The Swarm Plays process." The classroom became whimsical, filled with false starts, and invigorating (Krause et al.).

In acknowledging the influence of swarm intelligence principles in my exploration, I also must recognize that the impact is mostly metaphorical and exploratory rather than scientific. There is scientific backing for the idea of humans behaving like a swarm and the capacity for humans to solve problems collectively better than the highest-performing individual in the group. Krause et al. further express the idea of swarm intelligence this way, "Two or more individuals independently, or at least partially independently, acquire information and these different packages of information are combined and processed through social interaction, which provides a solution to a cognitive problem in a way that cannot be implemented by isolated individuals." (Krause, Ruxton and Krause 29) A collection of research continues to explore the capacities and characteristics of human swarm intelligence. This research is mostly contained to collaborative problem-solving and not as often applied to the pedagogy of creativity.

Even with those caveats, the experience was a pedagogical success—students' writing became freer and more engaged. After that, I focused on how using "swarm-like" structures may further help students. I used organizational structures used in describing animal swarm behavior and improvisation structures[2], concentrating mostly on self-organization. The development and implementation of such exercises resulted in several students awarded prizes in writing competitions and a visible bump on SAT reading comprehension scores. Students also reported a greater sense of

---

[1] His work can be found at https://www.theswarmlab.com/.

[2] Rather than the use of games in the classroom, I concentrated on finding the structures the games accessed and applied them to exercises. The "Tryptic" exercise below is an example of using the game structure to spark further creativity.

ownership over their educational outcomes and an increased ability to take collaborative projects to completion. At least three of those explorations resulted in performances[3].

A synthesis of all the different digital practices began to reveal an initial set of swarm behavior principles that guided my digital experimentation. These principles coalesced around seven areas--six that, as part of swarm behavior, can be implemented in a biological environment and one that is easier to achieve in digital spaces.

- **Self-organization**: Swarm intelligence and behavior is at its core decentralized (Garnier, Gautrais and Theraulaz 3). A swarm may follow a leader, but the power inherent in those moments of individual leadership is temporary and variable. The collective decides when leadership is needed and when it must move away. The collective is always in a fluid state of re-organization, skewing any hardening of hierarchies. In my pedagogical experimentation, I have attempted to use anonymity (more on this on the last point of the list) in the exercises to simulate non-hierarchical self-organization since anonymity provides masking of distinct identities within the group.

- **Simple attraction/repulsion rules**: Swarm intelligence relies on elementary algorithms of attraction/repulsion between all swarm members. In my experimentation, I have used a series of "feeders" (attraction elements—short, many times random prompts) and "limitations" (repulsion elements—small, also random rules that curtail advance in a particular direction) to simulate these conditions. I have added some of those samples below.

- **A collective common goal**: Swarms teem around a single biological need (build a bridge, forage) This aspect of swarm behavior has been the hardest one to decentralize in pedagogical situations, and for now, I have used my position as an educator to gather the collective interests towards a goal

3 *The Education Project,* at https://maladren.com/the-education-project, and *I.D.* at https://youtu.be/4JpnaivKDUQ, collaborative creations within the MCVTS Theatre program. *Lysistrata 2.0* (photos at https://maladren.com/lysistrata-vs-the-patriarchy), collaboration with students at the New Jersey Institute of Technology.

and come up with a common objective. I need more work in this field to decentralize the decision-making process in finding the target of a pedagogical community that is feasible within our current educational structures. The group's goal solidifies the group identity, so the best pedagogical outcome relies on a clear individual buy-in to a final purpose.

- **A defined collective space**: Swarms function as discrete entities, with boundaries that can be mathematically represented. In my practice, I have attempted to emulate these boundaries by creating closed pedagogical spaces in the classroom--defined and protected from external influences, such as they exist in public social media spaces. To avoid the destructive tendencies typical of anonymous, broad group participation, the creative group must be closed, with all members known to each other (even though anonymous in the exercises) and, thus, subject to a common goal and a group identity.

- **Exploration/Exploitation periods**: Swarms spend an indeterminate amount of time in exploration. This expansion period may appear chaotic and ill-defined until the amount of information gathered by the collective reaches a tipping point, and the exploitation period begins when the information coalesces into a collective solution. In the digital exercises, we aimed to allow for the chaotic gathering of material--each anonymous participant prodding in different directions, with minimal interference by any authority figure and, in our anonymous settings, with no intervention of hierarchies within the community to pour bias into the interactions. As of now, I have not been able to allow for an organic tipping point into the collective creation (our academic structures still demand deadlines and finishing points), and part of creating a more extensive digital space for this type of human swarm digital creation is finding out the organic movement between these two periods.

- **Individual anonymity:** Individuals in the group remain anonymous during the collaboration. There is no cognitive debriefing, so participants are not required or expected to defend any of their contributions. An anonymous setup also allows for individuals to pursue lone pathways and to group as they wish.

2. Swarm Pedagogy: Examples

I will describe three of the tools I have used to explore these principles in simple and accessible digital spaces. I have placed these in *swarmcreations.org* for public explorations. The three exercises are Tryptic, Swarm-up, and The Swarm Plays.

Tryptic

The straightforward exercise tool I called "Tryptic" best illustrates my approach to implementing some swarm pedagogy principles. Tryptic developed over several iterations as an online, phone-centered tool for free association. My students and I used this tool to build monologues and poems in an anonymous, collaborative setting. The exercise broke up into three sections: the feeder, the individual output, and the collaboration. The feeders came quite often from improvisational games I adapted to writing situations. In the "feeder" version of this exercise, I allowed participants to enter anonymously into a Google Doc divided into three columns, two for prompts and the last one for the combined offer.[4] Participants could enter text in either prompt and the response column in any order they wished, from the anonymity of their phones. Students offered disparate sentences, usually declarative of an object or a state ("I am an orange cat"). As they saw those offerings pairing, students would come up with small lines that could be uttered in a universe that contained both prompts ("I am a Queen;" "I am an orange cat;" "Oh my queen, we have captured the Cheshire cat!").

Over time, I developed several "feeder" exercises—all of them kept the participants anonymous and were theme-less. All the feeders aimed to create a series of sentences with no theme, no censorship, and no knowledge of who was writing what.

Almost every single time I have done feeder exercises, participants have broken into laughter at the combinations, at the relevance of some of the prompts, at the sheer pleasure of finding patterns of language where there was none. Students felt in control of their prompts. After practicing feeder exercises for a while, their mutual interests and innate and inherited metaphors for the moment they were living in began to come to the forefront. Those improvised prompts gave them control of their voices in an organic, collaborative manner. Students were not thinking any longer about "what

[4] In improvisational games this exercise is often called "I am a Tree." Its origin is debated but the game is widely used as a warm-up.

they should write about" or about "write what you know." The anonymous digital games allowed them a window into their dreams and nightmares.

I have placed Tryptic in a simple website form that I use for workshops with different educational groups. In the website, the exercise divides into three parts: a feeder (or exploration) part, a solo writing, and a "swarming" (the exploitation, or, in our case, collective creation).

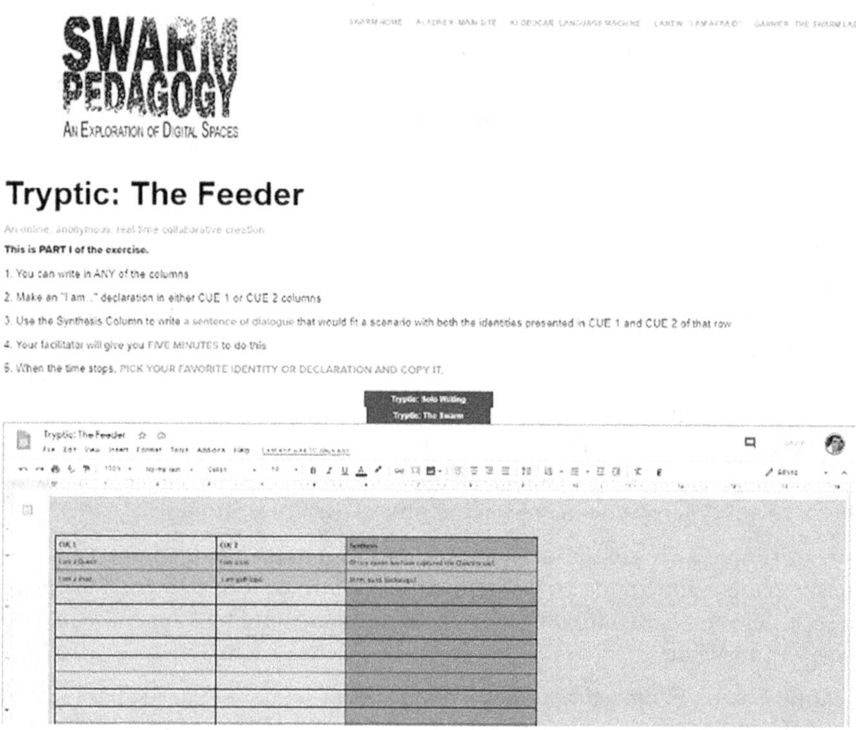

Figure 4.1. Tryptic's Feeder in the current website version.

The second phase of the exercise asks participants to choose one sentence from the feeder and begin writing as someone who could utter that sentence and use the chosen sentence at some point in the writing. Participants write individually, even though their words are visible to all:

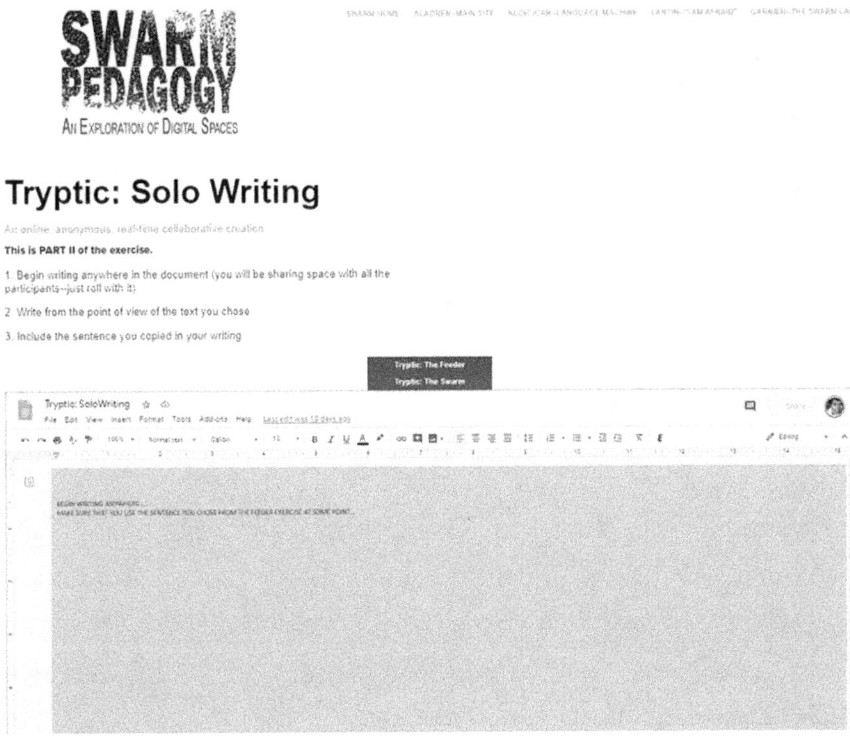

Figure 4.2. Solo Writing portion of Tryptic.

The last portion of Tryptic asks participants to combine all the paragraphs into a single collective creation. Participants are anonymous to each other, and they are allowed to edit, cut, copy, and paste any of the work in the document. Participants are encouraged to work with the impulses they see present if they feel so inclined and resist or reject if they feel like it. Because of their anonymity, no one in the group knows who is behind the group's actions.

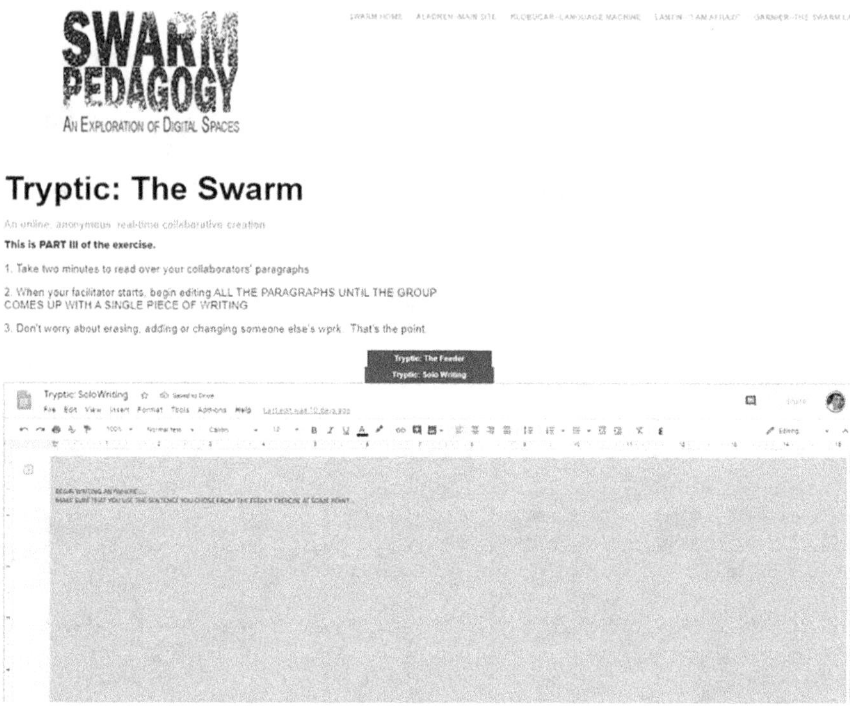

Figure 4.3. Swarming in the third part of Tryptic. [5]

Swarm-Up

Another simple exercise is "Swarm-Up," a visual warm-up tool. Swarm-up asks participants to create a visual collage organically, anonymously, and collaboratively. The purpose of this simple exercise is to allow the participants to begin an exploration of digital collaboration and to confront the natural fear of imposing on the ideas of others or having a judgment passed on their ideas. Because the exercise does not have a particular end product, participants begin to trust the process as the collaborative body exercises their collective imagination.

The exercise asks a group to create anything (I usually avoid any thematic guidance) as organically as possible. The activity may use readily available tools (such as Jamboard) to expand most classrooms' availability with an internet

[5] I have gathered some samples of material developed through this exercise at https://swarmcreations.com/small-swarm-samples.

connection. Ideally, members of the collective would be anonymous to each other. We achieved this through an incognito window in Google or using a copied link. Also, preferably, the exercise would be extended to a period beyond fifteen minutes to fully develop a collaborative space. I have enumerated below some of the instructions that I use to kick off the experience:

- This is just an exercise. We are not trying to create meaning or find metaphorical patterns that may be present in the group. Just. Have. Fun.
- Do notice if you are censoring yourselves or if you feel you are taking a risk. Do not fight it but notice it.
- Keep doing. Let the action get ahead of the thought.
- Respond, react, echo, erase, mirror, contradict. Play with others. Avoid creating your solo corner in the big playground and staying there.
- Use the tools--Jamboard is very simple. Play with the toys.

Beginning work is tentative and child-like, but the exercise itself allows participants to confront the fears of sharing a digital space anonymously.

Figure 4.4. First attempt at Swarm-up by a group of five adults.

The Swarm Plays

My third example deals with structures created to generate performing text. The most elaborate version of these exercises asked students to work over months on a single collaborative document under a set of anonymizing rules. Those included:

- Participants who did not know who was working on a document at any given time.
- A text that could be modified over weeks according to a series of "obstacles" the community had proposed. Those obstacles were assigned randomly.
- A community that created limitations on the number of words added or deleted in any given period.

In its essential structure, participants generated an open-source document from a variety of "feeder" exercises[6]. Then, the community established a set of arbitrary rules and prompts that allowed writers to input, edit or delete material randomly. As this process developed, participants agreed never to discuss ideas, plot, theme, or concept outside of the digital space, thus minimizing social hierarchies and pressures. Eventually, the community devised ways to make participatory input anonymous. The process (the exploratory phase) went on for months[7]. After that, we "harvested" or "swarmed" the material (the exploitive phase). In the last step, we include a "non-swarm" editorial process to abide by educational censorship of performance.

Some of the material made its way to performances. Much of the content went unused, but students' writing improved dramatically--it became daring, and their voices became more apparent and more assured.

3. A more equitable and approachable digital pedagogy

The results of those digital explorations began to coalesce around an improved ability to build spaces of collective creativity—which showed a

---

[6] https://swarmcreations.com/sample-feeders for a feeder exercise sample.

[7] https://swarmcreations.com/sample-rules for a sample day of rules and https://swarmcreations.com/sample-limitations for sample limitations. The rules and limitations kept changing as the process went on.

promise to provide students with the flexibility to adapt to changing systems and the social capital to organize as groups with peers.

In the late 20th century, creativity research refocused from individual traits to environmental traits surrounding individuals. Our western society had long stereotyped creativity as a trait belonging to a specific personality—the creative type (Dumas and Dunbar). We assumed that creativity was inherent, and training or improving it was an individual and challenging process. Stereotypes of the "creative persona" began changing in the 70s. Mihaly Csikszentmihalyi synthesized such refocus of creativity research when he stated that, "we need to abandon the Ptolemaic view of creativity, in which the person is at the center of everything, for a more Copernican model in which the person is part of a system of mutual influences and information" (Csikszentmihalyi 58). His systems model shifted the emphasis from creativity's internal cognitive process to the social and environmental process. Thus, the exploration slowly began to include the effect of the collective on such a process.

Once creativity research began to consider group dynamics, those dynamics became relevant in the pedagogy of creativity. The education field was also shifting towards an understanding of group dynamics in pedagogy--and the inherent power structures hidden in our teaching. As Paulo Friere stated in his seminal piece *The Pedagogy of the Oppressed:*

> The conflict lies in the choice between [...] being spectators or actors; between acting or having the illusion of acting through the action of the oppressors; between speaking out or being silent, castrated in their power to create and re-create, in their power to transform the world. This is the tragic dilemma of the oppressed which their education must take into account. (Friere 46)

In his view, until both teachers and students are "subjects," education remains an imposition--a violence, and creativity falters. The work that followed Friere further researched the imbalance in the classroom's power dynamics and how it continued to favor white, patriarchal, heteronormative structures and limited the education of more disadvantaged and vulnerable demographics.[8]

---

[8] bell hook's *Teaching to Transgress:* Education as the practice of Freedom, and Bettina L. Love's *We Want To Do More Than Survive: Abolitionist Teaching and the Pursuit of Educational Freedom,* among others.

Both these developments impact our understanding of the human creative process, with its reliance, in part, on tolerance and acceptance of wrong solutions and unbecoming outputs. The impulse to embrace those failures does not come naturally to vulnerable student populations, who feel the pressure to succeed, nor, in general, to the post-pubescent human. By the time we enter our teens, we have come to the life-long realization that our successes build our reputation within communities. In short, a large part of our student population fears backlash from peers and instructors if they do not conform to a set societal expectation of creative output.

Even in inherently creative situations (i.e., performing arts, creative writing), students consistently react to creative prompts with hesitation and fear—they want to know how to do it right. Those fears may stem from the authority the education system confers on teachers and the internal power dynamics in student groups. As educators or providers of creative spaces, our implicit and explicit bias may also exacerbate those fears. These power structures limit students' capacity to navigate perceived failure when engaging in the creative process—both analog and digital, and this danger manifests itself as self-censorship. The risk is not in technically subpar output, but output that becomes tentative, compliant, reactionary, and boring.

4. Creativity in digital spaces

Our current digital explosion exacerbates the problem. Our digital selves want to portray, at all costs, an image of success and invincibility. We want our every attempt at problem-solving or artistic output to be "liked." Amber Case made that point in her TEDtalk when she said that,

> You have a second self. Whether you like it or not, you're starting to show up online, and people are interacting with your second self when you're not there And so you have to be careful about leaving your front line open, which is basically your Facebook wall, so that people don't write on it in the middle of the night -- because it's very much the equivalent. And suddenly we have to start to maintain our second self. You have to present yourself in digital life in a similar way that you would in your analog life. So, in the same way that you wake up, take a shower and get dressed, you have to learn to do that for your digital self. And the problem is that a lot of people now, especially adolescents, have to go through two adolescences. (Case)

Youth now have, for the most part, a digital identity that is as important, vulnerable, and complex to them as their physical identities. The current educational policies have hesitated to acknowledge this fact and resist exploring it, leaving new generations navigating solo this uncharted territory. Current education policies treat our digital world as another tool—like a copy machine or a typewriter. However, students' relationship with the digital world was radically different. Those policies are a disservice to the new generation. Even though educators (and the educational institutions) are mostly not "digital natives," we need to develop pedagogical methods *inhabiting and participating* in the digital space, not just methods that *use* the digital tools. And we need to take those spaces seriously.

5. Conclusion

Since Paulo Freire's book, teachers have developed many techniques to disrupt the inherent power structures present in the classroom, to help students tolerate failure and diminish its impact in pedagogical groups, and, thus, to "free" the student to question, fail, and create. We must now look for ways to liberate students from oppressing powers developing in our digital world, and I propose that the way to do that is not by dismissing, banning, or diminishing this brave new world, but by inhabiting it, questioning it, and finding the most equitable uses for our students.

These explorations into anonymous digital spaces are still ongoing, and therefore conclusions are limited. I have encountered, though, enough encouraging developments to continue to create, develop and strengthen digital spaces that allow small communities to create collaboratively outside of the established--and many times, toxic--power dynamics that so often slow down or negate creative impulses. I hope to continue to develop models and methods that will allow us to create freely as digital communities and that, in the process, we can teach a new generation how to achieve digital identities that are creative, communal, and free.

In many ways, the digital landscape is terra nova. As such, it houses many of the exploring and exploiting impulses that have marked our previous human excursions. These next generations—cyborgs as they are already—are colonizing those unknown landscapes. Those with capital will take the lion's share of the resources, and those without will continue to serve purposes in those colonizations to adapt to the masters' orders. If our pedagogy is to be more than a reproduction of societies' structures, educators have to prepare students for the excursion into these lands if it is genuinely to be transformative.

This chapter is more call than praxis, more raising my voice to the chorus that is already sounding an alarm. If our society is liberating itself from algorithmic tasks and passing those burdens to the machine—as we have done since the beginning of our species—we need to rewrite the time usage of humanity towards creativity, collaboration, experiences, community, and we need to teach those values and skills in a new format, with new structures. Let us avoid the digital oppression (humans as data products) that may befall us if we do not pay attention.

Works Cited

Anyon, Jean. "Social Class and School Knowlege." *Curriculum Inquiry* (1981): 3-42.

Case, Amber. *We Are All Cyborgs Now.* n.d. <www.ted.com/talks/amber_case_we_are_all_cyborgs_now?language=en.>.

Csikszentmihalyi, Mihaly. *The Systems Model of Creativity: The Collected works of Mihaly Csikszentmihalyi.* New York: Springer, 2014.

Dumas, Denis and Kevin N. Dunbar. "The Creative Stereotype Effect." *PLOS One* 11.2 (2016).

Friere, Paulo. *Pedagogy of the Oppressed.* Continuum, 2005.

Garnier, Simon, Jaques Gautrais and Guy Theraulaz. "The Biological Principles of Swarm Intelligence." *Swarm Intelligence* 1.1 (2007): 3–31.

Jandric, Petar. *Learning in the Age of Digital Reason.* Rotterdam: Sense, 2017.

Krause, Jens, Graeme D. Ruxton and Stefan Krause. "Swarm Intelligence in Animals and Humans." *Trends in Ecology and Evolution* 25.1 (2009): 28-34.

Chapter 5

Maker Generation? The Uncertain Future of Students as Interactive Storytellers

Anastasia Salter

University of Central Florida

Abstract: In the wake of 2020's initial COVID-19 closures and widespread shifts to online teaching, our relationship with technology in the classroom is fundamentally changed. While questions of equity of access and availability of computing linger unresolved over this new distant learning, the increased ubiquity of laptops, tablets, and phones as part of higher education is inescapable. Yet this increased availability of the tools for making and sharing digital stories does not coincide with increased critical making across disciplines. Much of the experimentation discussed in this collection remains associated with computer science, digital media, or at best digital humanities: all disciplines that come with histories of exclusionary discourse and gatekeeping, as well as assumptions about what "making" looks like, and whose making is valued. How might we acknowledge our complex relationship with technology in the classroom, while striving towards transdisciplinary humanist pedagogy that empowers students as creative, critical makers?

Keywords: Pedagogy, Pandemic, COVID-19, Social Media, Digital Storytelling, Bots

***

As I write these words in the summer of 2020, higher education is preparing for a fall like no other on record: every day, universities are announcing their intention to move the semester online. Faculty, staff, and administration are trapped between apparently unresolvable competing interests of campus finances and safety; of accessibility and feasibility; of active learning and social distancing. The moment is rightly drawing our attention to practices of campus technology: who has access? Where are digital platforms fully integrated and valued? Prior to this unusual moment, admittedly the increasing presence of technology on the campuses of higher education seemed impossible to escape, from growing numbers of computers labs to students' own ubiquitous laptops, tablets, and phones at every class meeting. Sherry Turkle has been particularly critical of the role these devices play in moving us away from conversation (Turkle), an argument echoing in the demands to return to the classroom for meaningful learning, repeated in every op-ed that speaks out against laptops in the classroom. The rise of Web 2.0 brought an increased interest in these possibilities as well as several misconceptions, including the myth of the digital native and corresponding assumptions of student's inherent digital awareness that have proven far more complicated than originally imagined (Margaryan et al.). Yet such concerns reflect another, deeply embedded, binary that still harms humanities classrooms: the idea that these devices are for the reception of knowledge, not the making. Despite the fact that tools of making and sharing stories are likewise becoming more available and shifting not only into the hands of students but into the reach of educators across disciplines, much of the experimentation discussed in this collection is associated with computer science, digital media, or at best digital humanities—it is still far from common in the programs that need it most. Our complex relationship with technology in the classroom today must be mindful of these critiques, while pushing pedagogy further towards embracing the potential for students as makers both of things and of meaning.

The meaningful integration of new platforms for making into the classroom has been hindered in part by rates of adoption, which have been slowed by suspicion and even outright hostility towards new media in the classroom. Criticisms of technology in the classroom often center on the impact that technology has on attention, which is based on a flawed premise. Anyone who believes students are less attentive simply thanks to the presence of an iPhone or laptop clearly missed the number of students hiding books and magazines in their notebooks (I was always partial to trade fantasy novels, which were easy to hide under a desk). The failure of humanist higher education to fully embrace making through technology in the classroom is a failure of imagination, not merely of policy. Prohibitions on laptops and phone use reflect a fundamental insecurity, perhaps rightly—the suspicion that students will be more engaged by the continuous feeds of news and moments available at a touch from any of these devices than by a droning lecture is as well-founded now as it was with any available distraction in classrooms prior.

Pedagogy is continually in the process of being re-invented to accommodate the latest devices and technologies, but perhaps none has occupied as much collective attention as the iPad. Critiques of the iPad as a technology that encourages passivity on the part of its users are certainly valid: it is the ultimate example of a closed platform, physically shut to hinder inspection or modification, and gated in its applications review policies. Thus, the iPad has become emblematic of a type of technology use that implies mindless consumption. These critiques often focus on the patterns of consumption and learning that fragmented social media experiences suggest: this type of critique takes on new teeth in the wake of the 2016 presidential election, and again in the time of COVID-19, as questions of social media, knowledgeable media consumption, and the impact of fake news on voting outcomes and mask-wearing have been thrust into the spotlight. The only way to address these concerns is head-on, with the integration of critical making and a knowledgeable understanding of the algorithmic, procedural spaces of social media an essential goal for any humanist discipline.

1. Mapping Existing Social Media Making in the Humanist Classroom

Laptops and tablets have easily been integrated into higher education for more efficient completion and delivery of student work, as well as the distribution and consumption of learning materials. Even the most stalwart of professors in enforcing cell phone and laptop bans relies on technology constantly for every part of the higher education workflow, even if much of higher education is stuck in the remediation phase, replicating the traditional structures of the classroom for new methods of delivery. And with the widespread transitions resulting from COVID-19, these practices (and the hypocrisy underlying many choices in educational technology) are rightfully under increased scrutiny. This focus on consumption leads to several delivery-focused initiatives that have come and gone out of favor in the past years of ed-tech "innovation": iPad-enhanced textbooks, online lectures, MOOCs, and so forth. Most classroom attempts to harness students' existing interests in making and sharing content over social and media focus on Facebook and Twitter. This is more of a reflection of the interests of the professors than of students, as demographically, those two networks have a strong base of older, college-educated users (Duggan). Currently, several platforms represent continual types of making by students that often go unnoticed in the classroom:

- **Snapchat (launched 2011):** Snapchat is perhaps best-known for its combination of ephemeral messaging (messages sent on the platform are not saved unless the user actively decides to store them) and algorithmic filters (which range from offensive racial stereotypes to strange gags of vomiting rainbows. Snapchat offers an intimacy of communication that makes it a perfect testing ground for digital communication, and offers mostly untapped potential for pedagogical use (Ernstberger and Venable). TikTok (launched 2016) deserves similar mention and attention, although its continued availability in the US is currently in question.

- **Reddit (launched 2005):** While Reddit is rather notorious as a haven of trolling and offensive memes (Massanari), it is also an active site of community organization and information exchange of the type that any online classroom must envy. Reddit's voting-based system means that students wishing to share or create content on the network are continually subject to peer monitoring and audience evaluation (Gilbert). The conversations Reddit already enables put most online classroom discussion boards to shame.

- **Tumblr (launched 2007):** As a hub of fan community sharing, Tumblr is different from many of its compatriots. A range of content thrives on the platform (including amateur pornography, fanfiction, mash-ups, comics, selfies, etc.), making it difficult for newcomers to navigate. Despite the challenges it presents, Tumblr has been incorporated into classrooms for several assignments, particularly in writing (McGuire).

These are just some of the largest market-share platforms representing a few of the greater trends. Colin Lankshear and Michele Knobel argue for the importance of using the plural term "digital literacies" to discuss the range of conceptions of digital literacy that deal with both conceptual and operational levels of working with technology. The very range of social media platforms that students already participate in is a testament to the complexity of identifying and navigating digital literacies—and to the challenges inherent in making students better aware of how.

There are several levels of making in social media ecosystems that students can occupy. The roles typically referenced in higher education and used in the context of digital pedagogy include:

- **Producer:** Producers are what is typically meant by makers in the classroom: producers make the new content that circulates through the web. There's a vast range of types of producers and levels of digital literacy required for making content. Higher education initiatives are typically occupied with either students as writers or students as makers, but the range of making advocated in the classroom is far narrower than that of producers within social media.

- **Remixer:** Remixers (drawing on Lessig's concept of Remix culture) primarily take material from existing sources and reinvent it through mash-ups, modification, and other methods of combination. Successful remixing can require high levels of media literacy, and even move into the territory of production, particularly in communities dedicated to transformative works.

- **Amplifier:** Amplifiers primarily share, spread, or retweet content created by others. Ref Jenkins on Spreadable Media: amplifiers do the spreading, not the creating. Much of the original work of amplifiers is done through comments, tagging, or other metatextual acts rather than changing the text itself. Amplifying requires minimal media literacy.

- **Curator:** Curators sift through multiple sources and create a new destination that may hold a mix of original and found content. Curators are similar to amplifiers, but usually have more intention –they might host a site dedicated to Supernatural GIF organization by theme, for instance, or they might take humorous images and recaption them to fit a topic or commentary. Curatorial success is often judged by the number of people who follow or subscribe to the curated content.

Within these spaces, most of our students occupy the roles of curator or amplifier within social media with relative ease. Our coursework likewise tends to ask them to continue in those roles, rather than pushing them to develop a stronger relationship with the form—and in doing so, to better understand the pitfalls and challenges of algorithmically-driven platforms. This does not mean they understand the consequences or inner workings of the networks on which they operate. In this, then, social media offers a gateway to discuss our students as makers and storytellers on a visible level—any curator of a personal Facebook page is continually engaged in acts of storytelling and self-creation, performed for an audience. Elisabeth Buck's survey of students suggests: "students already intuitively negotiate and define audience, exigencies, and constraints when they decide whether something is, for instance, "Facebook-worthy" or "Twitter-worthy"—they simply need the critical vocabulary and distance with which to analyze it" (Buck). Like Twitter and Facebook, and their numerous offshoots and competitors, have become the definitive platforms for shaping the narrative of our times, this lack of literacy—and a corresponding disengagement with the inner workings of any platform—is a major lack in traditional humanist approaches to education, and one that can only be remedied with active making and deconstruction of those same platforms.

2. Examination through Critical Making

Drawing students into the level of producer or maker requires a more thoughtful integration of projects into pedagogy. Digital making that makes meaningful use of technology is recognizable by one central quality: it "includes projects that are interactive, media-rich, and, most importantly, cannot be (as) meaningfully composed or published in nondigital form" (Delagrange). Much of what is produced and circulated through social media is not inherently digital in its delivery: memes could be printed, tweets bound in a book, and Instagram photos committed to celluloid if desired. However, the interactive elements surrounding each of these individual acts, and the stories they are used to tell, start to elevate even quick acts of remixing and production towards more meaningful making. Digital storytelling takes this one step further, with works of interactive fiction, games, hypertext, and other forms defying easy conversion to non-digital modalities. Digital storytelling is a form of critical making focused on meaning-making. Often, it begins with fragments, whether those are tweets gathered in a Storify or nodes connected in a hypertext in Twine. The weaving of connections is as important to the content to creating the story.

However, the form of the act is not as important to learning digital literacy as its context. When making (digital or otherwise) moves into the classroom, it can be intentionally scaffolded with principles of literacies, rhetoric, and communication. Daniel Chamberlain defines work grounded in this way as critical making:

> At a deeper level, an emphasis on robust forms of critical making includes a process of production, reflection, reformulation, and reconsideration, or making as a way to ask better questions. For students who are not fully persuaded that academic papers make a difference in the world, critical making projects often open an

> avenue to critical intervention, and help them understand that making is something done by authors, scholars, and audiences, and that students themselves may occupy these roles." Critical making is often discussed in the context of scholarship, as an alternative to the traditional essay, a pedagogical standby continually subject to critique for its odd lack of audience and archaic format. However, critical making extends beyond critique into artistry: in making, design and function are not separate. The message (or story) of a work is intertwined throughout its making" (Chamberlain).

Chamberlain describes an alternative to the tradition of the academic paper grounded in digital literacies, that harnesses the production elements of students' experiences towards critical thinking and meaning-making. This places a high demand on us as instructors looking to instill critical making practices in our students: it is not enough to master a tool or software program. A critical maker reflects on the tool itself and rejects, supplements, extends, and critiques it as part of the process of making. Consider for instance one of the most perfectly reflexive examples of critical making, *Phone Story* (Molleindustria 2011). Initially released for iOS, *Phone Story* was banned for telling the hardware production story of exploited and inhumanely treated laborers in the factories and mines behind the making of iOS as a device (*Phone Story - Apps on Google Play*). The app not only used the phone as a platform, but drew attention to it, demanding the player to be aware of the realities of what they were holding. The app was unsurprisingly banned from iOS, although Android's laxer marketplace allowed it to remain.

Phone Story makes for a compelling example of an educational game because it both embraces and rejects the platform of its construction and consumption. This critical approach to any ecosystem, hardware or software, is an important part of critical making that goes beyond the tools-based technical knowledge that is often associated with the practice. Stolley is particularly critical of tools-focused rhetoric in education: "It is alarming when software, commercial or otherwise, comes to signify entire digital genres. Compare the number of results in a search engine for PowerPoint best practices versus slideshow best practices. The results suggest that vendor lock-in has as much of a grip on how people talk about production as it does on what actually gets produced" (Stolley). Stolley's charge to think outside of a particular software program gets to the crux of integrating making into education. Making in the classroom is not about teaching a platform or interface: it is not accomplished through requiring students to "be" on Tumblr, or even adding lessons on the use of a storytelling program such as Twine or Inklewriter. Instead, it is about creating awareness of how those platforms shape making: death by Powerpoint is ubiquitous because of how Powerpoint is made.

Courses that engage students in critical making break down pedagogical expectations by rejecting the knowledge economy for participatory culture: "a course that approximates a participatory culture emphasizes experiential learning by having students engage directly in the processes relevant to the course topic rather than primarily consume course materials that explain them. By focusing on processes, knowledge is then understood as ways of thinking and making rather than information or facts that are today readily available, and even debated, without classrooms of higher learning" (Mair). This does not decrease the importance of "consuming" knowledge: instead, it changes the valance, suggesting that the act of consuming must be done with a critical, reshaping gaze even within the context of the classroom. This critical gaze even includes a potential rejection of the constraints and assumptions of software tools, social media sites, and other technologies.

Roger Whitson extends Chamberlain's definition of critical making: "Critical making is, therefore, not simply a way of understanding history, but is also an investigation of what could have been, what Trevor Owens calls alternative historicism…the making and rewiring of past technologies provides us with an opportunity to see how, for example, colonialism might have played out differently with a different set of economic, sociological, or technological circumstances" (Whitson). It is no surprise that what Whitson described is not unlike the work of storytelling itself, as he situates it in an examination of steampunk as a genre. It is particularly significant that this approach to critical making is literary rather than procedural. Media literacy is as fundamental as literacy to the functioning of our society. The division of majors into STEM fields, where computers are generally embraced, and humanities fields, where computers are viewed as the subject of a not-so-new "digital" humanities rather than an integral part of every act—is artifice at best and harmful at worst. Yet debates surrounding the value of digital making are often centered on code. Barack Obama's administration orchestrated initiatives such as "Hour of Code," and pushed for coding to become a fundamental part of the curriculum. Douglas Rushkoff's *Program or be Programmed* casts the matter more darkly, suggesting that those that do not learn to code are perpetually at the mercy of the algorithm (Rushkoff). There's some truth to that: a study of Facebook users suggests that despite widespread coverage of Facebook's news feed filtering, many users still don't know that their experience is governed by algorithms (Eslami et al.).Given the potential impact of the filter bubble and algorithmic amplification of news on the 2016 American presidential election, and by all appearances, the 2020 election, this illiteracy presents renewed cause for concern and educational intervention (Tufekci).

However, starting the conversation about critical making in the classroom at the level of code can be alienating and counter-productive. Despite the still somehow entrenched myth of the digital native, code literacy is not a fundamental part of education, and the mathematical knowledge required to engage in algorithmic literacy makes it inaccessible to many thanks to the low emphasis on math education. Furthermore, the pace of technology is daunting to classroom instructors, who often have their own disciplines to contend with before keeping up with changing interfaces and standards. The additional demand of code literacy to constitute a maker-friendly space is even more unrealistic and can serve more as gate-keeping than to promote meaningful engagement. Not all students will go on to develop software or build tools for making, but all students will use digital tools, and they must be prepared to demand more and understand how the tools are in turn shaping them.

Karl Stolley offers a manifesto for "lo-fi" production in courses focused on making, suggesting that removing the abstraction of the graphical interface allows for an investigation on a different level:

> The problem is that the market for GUI [graphical user interface]-driven software has conditioned people to expect the same ease everywhere. To be sure, there should be no need to write source code just to check email or to book a restaurant reservation online. But that does not mean that no occasions exist when writing source code is absolutely necessary. A lo-fi approach rejects intuition and reflex in exchange for the uncomfortable uncertainty and time-consuming struggles of research and learning. Intuition and reflex are only for today, for the person making something. What is being made, and for whom, are always different problems, project to project. Research-driven lo-fi production is not just about investigating the how of production that visual interfaces embed and make intuitive. Lo-fi production directly addresses essential audience-driven concerns of digital creation that are also the most stable and sustainable: the what and the why, under the human and technological constraints of for whom (Stolley).

This lo-fi approach is enticing for both its ease of incorporation into the classroom and its potential to produce lasting works: as any former devotee of Adobe Flash knows, the latest platform quickly becomes an unsupported, abandoned program. However, it does ignore some of the opportunities inherent in spending time making with as many programs and networks as possible. In our personal use of social media, we tend to favor the platforms we are comfortable with: higher education offers opportunities to push students beyond comfortable roles (as amplifiers and curators) and familiar networks to making new experiences that serve as critique and reflection.

Digital storytelling is based on principles that extend beyond interfaces and drawing students into the process of constructing (and dismantling) digital stories may be one of the most important goals facing us in an era of demonstrably decreased media literacy. The fundamental literacy of maker culture is not in assemblage but in deconstruction. With the awareness that systems are constructed and intentionally comes the ability to challenge the underlying assumptions and biases of any platform. The current generation is entrenched in making, spread continuously to new audiences through social media. Educators have the opportunity and responsibility to extend that making, building upon it with opportunities to embrace (not ban) technology in the classroom. Only through these reflective practices can we ensure a future of critical making, building on critical thinking to challenge and rethink both the social and technical structures surrounding us.

3. Making Bots to Understand Bots

Procedural writing is a particularly timely intervention for such efforts at forefronting critical making. While most generative writing (including many of the exemplars addressed in this book) has barriers to creation that are beyond the average humanities classroom use case, a few approaches have stood out for their replicability. For instance, Nick Montfort's *Taroko Gorge* has inspired so much imitation that it was featured in the third volume of the *Electronic Literature Collection* as a site of newcomer experimentation: however, most such usage is simple replacement through substituting words for the original poem to change the meaning (Boluk et al.). The leap from such a text to a more nuanced generation of a similar JavaScript program can be intimidatingly daunting, although works such as Montfort's own on programming for the humanities offer strategies for bridging that gap (Montfort). However, for a more accessible critical making tool, JavaScript library Tracery stands out as an exemplar: it was developed by Kate Compton as "an authoring tool for casual users to build grammars to generate interesting stories. This emphasis on the casual user is unusual in procedural tools of any kind. Kate Compton's own aesthetic puts an emphasis on the feminine and the playful, inviting the user into a style not typically associated with the space of computer programming.

Tracery breaks the making of procedural texts into a grammar of JSON, or Java Script Object Notation, allowing for easy substitutions as well as more sophisticated patterns. For example, here's a simple sentence generator from a Tracery tutorial aimed at historians (Graham):

```
{
        "origin": [
                "I #being# #placewhere#"
        ],
        "being": [
                "am so lonely",
                "am so lonely",
                "am begging you",
                "am turned towards the sun"
        ],
        "placewhere": [
                "in a pot",
                "in a windowsill",
                "fallen over"
        ]
}
```

Note the readability: while the syntax is somewhat demanding, the fundamental logic of identifying categories and providing replacements to fill an "origin" sentence structure is fairly straightforward, and easy to expand upon without any knowledge of the code by which the grammar is interpreted. Making through Tracery is thus one of the most accessible exercises available for playing with the affordances of generative text: its patterns can be rapidly brought to social media platforms thanks to "Cheap Bots, Done Quick!," a tool created by George Buckenham to allow the deployment of Tracery-powered bots to Twitter (Buckenham).

Tracery-powered bots are part of a landscape of humanist resistance and discourse that exemplifies the potential of critical making. Mark Sample's NRA protest bot demonstrates the potential of Twitter bots to amplify and critique: as he describes in his article on the bot's making, "I turned toward algorithmic culture, to the margins of the computational world. I created a bot of consolation and conviction that — to paraphrase Phil Ochs in 'When I'm Gone' — tweets louder than the guns. The bot I made is @NRA_Tally. It posts imagined headlines about mass shootings, followed by a fictionalized but believable response from the NRA. The bot is topical, grievously so. More critically, you cannot mistake it for bullshit. The bot is data-driven, populated with statistics from a database of over thirty years of mass shootings in the U.S." (Sample). Other examples of bot-making as art and resistance fill the "Bots" section of the third *Electronic Literature Collection,* including Darius

Kazemi's "Two Headlines," which he frames with the statement: "Two Headlines is a Twitter bot that attempts to automate a kind of lazy Twitter joke where a human confuses the subjects of two news items that everyone is talking about on Twitter. An unintended consequence of its particular algorithm is that the bot that also writes near-future late-capitalist dystopian microfiction, in a world where there is no discernible difference between corporations, nations, sports teams, brands, and celebrities" (Boluk et al.).This type of making disrupts consumption—it demands reflection, and engagement, with the very making of the headlines the algorithm reimagines.

4. Making, Reframed

It is impossible, writing this in the summer of 2020, to predict what's next for the humanist classroom—or the United States. At the moment of this chapter's writing (only a few weeks out from the start of classes in Fall 2020), the *Chronicle of Higher Education* features a split pie graph, with only around 23% of colleges planning on an in-person reopening while the rest hover around indeterminate, hybrid, or fully online plans (Staff). Larger questions of the economic feasibility of humanist education have already been circling, like vultures, picking at already desiccated humanities departments in public universities suffering from a new blow in an ongoing funding crisis. Students are surrounded by misleading, contradictory, information, and the push against humanities education is escalating, predictably, alongside the economic crisis. The tools of accessible, unintimidating, critical making are more important than ever—they do not demand financial investment from students in complex software, but offer the rewards of new insights and new stories to push back against the overwhelming tides of fake news. The divisions of students by major into STEM fields, where computers are generally embraced, and humanities fields, where computers are viewed as the subject of a not-so-new "digital" humanities rather than an integral part of every act—is artifice at best and harmful at worst. Interactive storytelling offers an intervention that allows us to break down these assumptions, and in doing so, imagine more inclusive, collaborative approaches to interactive storytelling. Such tools can provide a gateway into more intensive algorithmic making (such as some of those examined later in this book, such as VR), but also can serve on its own as a tool for reimagining the classroom at a moment when such reinvention is urgently required.

Works Cited

Boluk, Stephanie, et al. "Electronic Literature Collection - Volume 3." *Electronic Literature Organization*, Feb. 2016, http://collection.eliterature.org/3/.

Buck, Elisabeth H. "Facebook, Instagram, and Twitter: Oh My!" *19.3*, Kairos: A Journal of Rhetoric, Technology, and Pedagogy, 15 May 2015. *kairos.technorhetoric.net*, http://kairos.technorhetoric.net/19.3/praxis/buck/index.html.

Buckenham, George. *Cheap Bots, Done Quick!* 8 Feb. 2017, https://cheapbots donequick.com/.

Chamberlain, Daniel. "Critical Making | Fieldguide." *Media Commons*, 13 May 2013. *mediacommons.org*, /fieldguide/question/what-are-differentiations-and-intersections-media-studies-and-digital-humanities/response-8.

Compton, Kate, et al. "Tracery: Approachable Story Grammar Authoring for Casual Users." *Seventh Intelligent Narrative Technologies Workshop*, 2014. *www.aaai.org*, https://www.aaai.org/ocs/index.php/INT/INT7/paper/view/9266.

Delagrange, Susan H. "When Revision Is Redesign: Key Questions for Digital Scholarship." *14.1*, Kairos: A Journal of Rhetoric, Technology, and Pedagogy, 15 Aug. 2009. *kairos.technorhetoric.net*, http://kairos.technorhetoric.net/14.1/inventio/delagrange/production.htm.

Duggan, Maeve. "Demographics of Social Media Users in 2015." *Pew Research Center: Internet, Science & Tech*, 19 Aug. 2015. *www.pewresearch.org*, https://www.pewresearch.org/internet/2015/08/19/the-demographics-of-social-media-users/.

Ernstberger, Jon, and Melissa A. Venable. "Can Snapchat Bridge the Communication Chasm in Online Courses?" *ELearn*, vol. 2016, no. 2, Feb. 2016. *02-01-2016*, doi:10.1145/2893467.2893353.

Eslami, Motahhare, et al. "'I Always Assumed That I Wasn't Really That Close to [Her]': Reasoning about Invisible Algorithms in News Feeds." *Proceedings of the 33rd Annual ACM Conference on Human Factors in Computing Systems*, Association for Computing Machinery, 2015, pp. 153–162. *ACM Digital Library*, doi:10.1145/2702123.2702556.

Gilbert, Eric. "Widespread Underprovision on Reddit." *Proceedings of the 2013 Conference on Computer Supported Cooperative Work*, Association for Computing Machinery, 2013, pp. 803–808. *ACM Digital Library*, doi:10.1145/2441776.2441866.

Graham, Shawn. "An Introduction to Twitterbots with Tracery." *Programming Historian*, Aug. 2017. *programminghistorian.org*, https://programminghistorian.org/en/lessons/intro-to-twitterbots.

Lankshear, Colin, and Michele Knobel. *Digital Literacies: Concepts, Policies and Practices*. Peter Lang, 2008.

Mair, Kimberly. "Participatory Culture and Distributed Expertise: Breaking Down Pedagogical Norms or Regulating Neoliberal Subjectivities?" *The Journal of Interactive Technology and Pedagogy*, 16 June 2016, https://jitp.commons.gc.cuny.edu/participatory-culture-and-distributed-expertise-breaking-down-pedagogical-norms-or-regulating-neoliberal-subjectivities/.

Margaryan, Anoush, et al. "Are Digital Natives a Myth or Reality? University Students' Use of Digital Technologies." *Computers & Education*, vol. 56, no. 2, Feb. 2011, pp. 429–40. *ScienceDirect*, doi:10.1016/j.compedu.2010.09.004.

Massanari, Adrienne. "#Gamergate and The Fappening: How Reddit's Algorithm, Governance, and Culture Support Toxic Technocultures." *New Media & Society*, vol. 19, no. 3, Mar. 2017, pp. 329–46. *SAGE Journals*, doi:10.1177/146144481 5608807.

McGuire, Meghan. "Reblogging as Writing: The Role of Tumblr in the Writing Classroom." *Engaging 21st Century Writers with Social Media*, IGI Global, 2017, pp. 116–31. *www.igi-global.com*, doi:10.4018/978-1-5225-0562-4.ch007.

Montfort, Nick. *Exploratory Programming for the Arts and Humanities.* MIT Press, 2016.

Phone Story - Apps on Google Play. play.google.com, https://play.google. com/ store/apps/details?id=air.org.molleindustria.phonestory2&hl=en. Accessed 31 July 2020.

Rushkoff, Douglas. *Program Or Be Programmed: Ten Commands for a Digital Age.* OR Books, 2010.

Sample, Mark. "A Protest Bot Is a Bot so Specific You Can't Mistake It for Bullshit." *Medium*, 4 Oct. 2015. *medium.com*, https://medium.com/@sam plereality/a-protest-bot-is-a-bot-so-specific-you-cant-mistake-it-for-bullsh it-90fe10b7fbaa.

Staff, Chronicle. "Here's Our New List of Colleges' Reopening Plans." *Chronicle of Higher Education*, 30 July 2020, https://www.chronicle.com/article/heres-a-list-of-colleges-plans-for-reopening-in-the-fall/.

Stolley, Karl. "The Lo-Fi Manifesto, v. 2.0." *20.2*, Kairos: A Journal of Rhetoric, Technology, and Pedagogy, 1 Jan. 2016. *kairos.technorhetoric.net*, http://kai ros.technorhetoric.net/20.2/inventio/stolley/index.html.

Tufekci, Zeynep. "Opinion | Mark Zuckerberg Is in Denial." *The New York Times*, 15 Nov. 2016. *NYTimes.com*, https://www.nytimes.com/2016/11/15/ opinion/mark-zuckerberg-is-in-denial.html.

Turkle, Sherry. *Reclaiming Conversation: The Power of Talk in a Digital Age.* Penguin, 2016.

Whitson, Roger. *#mla14: How to Theorize with a Hammer and an Oven: Steampunk & Critical Making.* 2014.

Chapter 6

Netprov: Collaborative, Online Roleplay as Art Form

Rob Wittig

University of Minnesota

Abstract: Bringing the terms "internet" and "improvisation" together, the term "Netprov" signifies an increasingly popular, creative use of social media, where fictional scenarios are devised and performed over a chosen network platform. For nearly a decade, common hosting programs like Twitter and Instagram have been commandeered by different troupes of writers, performers, students, and casual players role-playing together according to scenarios – fictional premises – which operate like the "rules" of childhood make believe. By answering the seemingly simple question: "how long should a netprov be?" electronic literature artist and theorist Rob Wittig provides an overview of netprov and a consideration of what netprov draws from the tributary fields of literature, theater, film, games and internet practices, before extending an invitation for further scholarship and play.

Keywords: Improvisation, Social Media, Performance, Multimedia Platforms

***

As more and more people come to my writing partner Mark Marino and me for advice about producing their netprovs (internet roleplay fictions) for the classroom or for a general audience of players, we get the question "how many days or weeks long should a netprov be?" We always have to take a deep breath before answering because the seemingly simple query opens up a host of complex and instructive issues about the shape, size and borders of a collaborative artwork in the digital age, as well as the nature of the co-creators' or the readers' experience in real time, the right of anyone to declare a netprov "done," and the design and maintenance of an archive of the project. This essay is a sketch of these parameters, a tentative answer, and a call for further scholarship and criticism.

But first, what is netprov? Combining the terms "internet" and "improvisation," the term "netprov" signifies an increasingly popular, creative use of social media, where fictional, often roleplayed narratives are performed serially in an available network platform. If your cat has a social media account and interacts with other cats — that is netprov. If you are among the nearly two million authors who contribute to the Facebook group called "A group where we all pretend to be ants in an ant colony" (or one of the fifteen-plus similar groups) — that is netprov. When somebody invents a make-believe event and posts about it in real time to crack up their friends — that is netprov. Netprov uses simple narrative rule sets, the same kind of rules that govern children's make-believe play and stage improvisation, as algorithms to guide the stories' development. What Mark Marino and I ask is: when informed by artisticstorytelling techniques from literature, theater, film and the web itself, can netprov become an art form?

Among the netprovs Mark Marino and I have produced in response to this question is *#1WkNoTech One Week, No Tech*, in which we all pretend to go for a week without the internet and tweet the heck out of it, posting selfies of our blissful walks in the woods, complaining about missing Netflix, and craving our daily minimum requirement of likes and retweets (2016). Another is *#fixurl8tionship Fix Your Relationship (For the Camera)*, in which we invited people to:

> Join a squad of Instagram influencers who "FIX" fans' relationships by helping them look*GREAT* for the camera despite how they feel inside!!
>
> And seek advice from another fixurl8tionship influencer about your character's failing relationship!
>
> Play a character you invent and improvise! (2018)

Mark and I offered two modes of participation in *#fixurl8tionship*:

> **Quick Play in Your Own Instagram Account**
> As a "fan" read and enjoy posts on the #fixurl8tionship hashtag! Like / Follow / Share!
> Ask an influencer a relationship question in comments, then
> When you receive the (usually awful) advice, report back on what happened in a comment!

> **Create Your Own Influencer Character!**
> If you want to play a flawless #fixurl8tionship influencer, there're a lot of great tips at the website, including our step-by-step method for creating a character account in Instagram. (2018)

An example of a more elaborate, month-long netprov was announced in a press release:

> Meanwhile Netprov Studio announces All-Time High, a participatory netprov (networked improv narrative) game that will take place on Twitter throughout July. Based on a concept by Claire Donato and Jeff T. Johnson, All-Time High was developed with Meanwhile Netprov Studio founders Mark C. Marino and Rob Wittig. (Marino and Wittig, "Press Release," 2015)

The program for the event followed on the website, *All-Time High: A Netprov*

> In All-Time High we find ourselves—our own high-school-aged selves—together at the same high school in July of 2015. What a nightmare, right?! And yet, what an opportunity. For comedy, if nothing else. And maybe even a bit of redemption.
>
> Readers become co-creators of the Twitter-based netprov with the use of the #ATH15 hashtag, and choose their own adventures at All-Time High. The official Twitter account, @alltimehigh2015, will operate as a PA system and make announcements over the course of the netprov.
>
> Throughout July, participants will play out the differences (generational, geographical, subcultural) and the commonalities (stress, sugar, hormones) of life that go on at one universal high-school. Using their own Twitter accounts or creating new ones, participants play their teenage selves and fictional or historical characters reimagined as high-school students or teachers.
>
> This hyperactive summer school is scheduled around four grand, archetypal high school events—The Talent Show, The Big Game, The Big Dance, Graduation—at which participants gather on Twitter for an hour each week at 11:00pm EDT (8:00pm PDT). From Thursday to Tuesday students, teachers, and PTA members will work up some good old-fashioned (and new-fangled) high-school drama while avoiding glare of the Vice Principal (@ATHvice). In an action-packed first week, The Talent Show shares its stage with The Big Test, and students rally to

> rename the decidedly dated team, The Fighting Cocks, and pick a new mascot. (2015)

As a netprov "netrunner" (a term drawn from the television term showrunner which indicates the lead creative person), I think of netprov as a common creative impulse that produces countless written and photographed fictions all around us all the time and has the potential to become a true art form. As a scholar, I have identified the following characteristics of the kinds of netprovs that particularly interest me: netprovs that

- Create fictional stories that are networked, improvised in real time, and often collaborative.
- Use multiple media platforms simultaneously.
- Are experienced as performances as they are published and are read later as literary archives.
- May incorporate breaking news during the performances.
- May include content that is topical and satirical.
- May use models or actors to physically enact characters in images, videos, and live performance.
- May include players who play, as actors, the characters they create and write.
- May require players to travel to certain locations to seek information, perform actions, and document their activities.
- Accommodate episodic and incomplete reading.

The platforms chosen for particular netprovs, such as Twitter and Instagram, bring along with them certain expectations for the behavior of their participant communities and about the outline and duration of the project and are, in their own ways, algorithms. But by stepping back and thinking of netprovs as art works instead of just epiphenomena of the platform, other cultural systems of expectation about art works brought into play. What we want to do here is draw from some of these other cultural systems, see what they have to offer to help define the limits of the netprov (what is part of the netprov and what isn't), and to describe their user experiences — all in the service of answering the question: how long should a netprov be?

I come from a literature background, so my expectations are always haunted by books with their fixed typography and visible and graspable size, which converts in my mind into an estimated duration of reading, short or long. As I see it, what is within the physical book is part of the work, what's not is not. But it's in the use pattern of books that the complexity lies.

Use pattern is a term I use in graphic design education to get students to really think about what the audience's ideal use of their creations would be. A poster announcing an event wants you to a) notice it, b) pause long enough to receive its invitation, and c) remember to attend the event. Once the event has passed, the poster wants you to keep it as an art object. A horserace betting ticket is a precious contract, a valuable financial instrument, that transforms into a piece of trash in the blink of an eye.

The actual use pattern of books is brutal. They are skimmed, read out of order, read when tired or drunk, cheated on as the reader has affairs with other books, shelved and displayed - unread - as backdrops to video conversations, and even abandoned completely and sold for pennies. It's important, as we consider the incomplete readings that most users make of netprovs, not to hold in our mind a false image of the supposed completeness and wholeness of the book *experience*.

Netprovs unfolding in real time also have many, but not all, characteristics in common with theater. Theatrical works are fictions roleplayed in real time, sometimes highly scripted, sometimes improvised. A scripted play lives at least three lives, as a text, as a performance, and potentially as a filmed archive, even though most professional theater contracts strictly forbid the sharing of recorded performances outside of the theater building and to anyone outside the company. In the theater tradition start times are fixed and strict, durations are commonly in the one-to-three-hour range, and a round of applause signals the end. The work, however, can be repeated an unlimited number of times, and each performance is considered a unique and irreplaceable event. Audiences are seated facing the work and generally know which people are part of the work and which are not. Performers are named and credited in a paper document.

Netprovs, like theatrical improv, consist of roleplaying that can be planned or improvised and can be repeated unlimited times with different results. Netprovs in social media may have launch dates but, being collaborative and crowdsourced, have no firm endings, and mostly just fade away gradually as the creators move on to other things. It also is never quite clear who is playing which characters – friends or strangers – and it is a particular creative pleasure of the form to create multiple accounts and steadfastly deny authorship of some of them.

But netprov also has formal elements modeled from film and TV. Movies and TV are essentially all archive, the live performance having been witnessed only by a few. Filmed theater – a movie – when shown in a theater, is used like a play. Episodic television used to have fixed weekly timetables and predictable season lengths. However, the use patterns of movies and TV have changed dramatically since the advent of streaming media. Crowdsourced and user-produced, YouTube was launched in 2005. Movie and TV industry outlets Netflix and Hulu were founded in 2007 and by 2013 Netflix was winning Emmy Awards for original content. The term "binge-watching," which describes watching entire seasons of episodic content in a matter of days points indistinctly at a more complex phenomenon. The instant availability of content seasons is new, but more significantly, the streaming services remember where you left off and allow you to start again at that point. This advance renders forms such as episode and season meaningless except as cultural conventions – holdovers of the technical limits of the past such as we see in the concept "books" (the amount of text that can fit on a scroll) of the Bible. Streaming media, instead of dictating time and place of use, are now built into the schedules of users at the users' will. The scope of use patterns is staggering, ranging from an academic colleague who writes essays with their text on one screen and the TV series *The Office* running beside it on the other, to sleeping with *Friends*, to driving with *Breaking Bad*.

Netprov, as Mark and I have formulated it, is intended primarily to be done at work, in little moments of respite from toil. Netprov represents a new use pattern of composition where, instead of the grant-funded months of uninterrupted writing in the cabin by the lake, creativity takes place in the weave of everyday life, more like running jokes among friends, more like short Instagram videos than Hollywood features. But this makes netprovs extraordinarily difficult to delineate since they are fragmented and woven into other streams of content, much in the way our streaming videos are cut into thousands of little chunks, sent in fragments through radically different pathways according to internet protocols, and then reassembled on our machines into a seemingly undivided flow. Only our vestigial mental constructs of "movie," "show," and "episode," define the limits of those works. If netprov is to draw anything from movies and TVs it is from their unconscious and mostly unacknowledged *actual* use than from their historical forms.

Besides drawing from literature, theater and film, netprovs are also a lot like games. In netprov there is usually a specified concept or fictional frame (e.g. we're going to pretend to go without technology but actually use it). There is a sense of play both in the sense of competitive play, a delimited arena in which the adventures are not as consequential as in real life, and of imaginative play or make-believe. The limits of a game can be thought of as, broadly, what happens within its rule set (charades) or upon its platform (Fortnite). Games, like netprov, are replicable, and with even more variation than traditional theater performances, variation-within-structure and learning curves that openly court binge play. I often use analogies to game play to invite newcomers to netprov in order to evade the unfortunate echoes of "hard homework" that many associate with writing because of their educational experiences. But here again, games are only of limited use in defining the limits and use pattern of netprov. We love it when people binge on netprov, but most netprov participants feel they have made a complete artistic statement in a particular project in a time frame a little longer than a week and a little shorter than a month. Games aim to keep players bingeing for years. Plus, many major games market new versions of the game every year effectively rendering the previous versions obsolete and playable only as nostalgia. This is very different from other art forms. Lady Gaga's *Star is Born* (2019) doesn't invalidate Judy Garland's (1954) or Freddie March's (1937).

Social media practices themselves contribute a lot to netprov in the way of use patterns. The weave, the feed of Twitter and Facebook provide most of the elements of the real-time netprov experience for our social media: instant publication, typography and formatting, reverse-chronological listing, names, avatars, and timestamps, links, threads and voting of various kinds. But the old, original feed of social media (displaying a user's selected authors in strict reverse chronological order), useful as an art form, has now, since Facebook's addition of the "like" button in 2007 and similar systems, been replaced by the "force feed," complex algorithms designed to amplify users' perceived desires and strategically insert advertising. We want to collaborate with other netprov players, not with the algorithms and their corporations who are determined to oversimplify us and make us predictable consumers. One of the delightful characteristics of art is that it can be as unpredictable as life is, as real human beings are. A key moment in social media design was the introduction of the infinite scroll around 2006. Before this innovation, it was possible to feel that you had seen every post from all your friends since the last time you had checked. You could scroll down until you hit posts you'd already read. Facebook and others, by monitoring users use patterns, detected that this was the point at which many people would leave the platform and go do other things. Since social media platforms share the same financial goal as the makers of slot machines — user time on device — they permanently opened the trap door at the bottom of the screen and developed ways to make the page, and hopefully the experience, endless.

One of the things that players of big, active, social media-based netprovs report is a troubling sense that they are always missing a few things, missing posts by others, or experiencing them in the wrong sequence. Some of this is due to the sheer speed that posts will appear in a hashtag during a very active netprov. But, sure enough, since the big platforms "customize" feeds to each consumer, players sitting side by side actually will see a slightly different version of the netprov. It is at this point that our encouragement to netprov players to be OK with episodic and incomplete reading comes into play. Mark and I remind them not to think of using netprov the way you think you should read a book (completely and diligently) but more like the way you actually read a book, with normal lapses of attention, skims and slides. We also share with them the reassuring dictum of netprov: "netprov means never having to say you're sorry." Netprov players frequently apologize for not having written more (shame created by education) and not having followed the flow closely enough (shame created by social media platforms themselves as they methodically stoke the fires of FOMO, fear of missing out). It's impossible to be a completist on these platforms, we say to them. Give yourself a break; play as much as you want to and have fun!

It is this very experiential incompleteness – not even its most active participants are sure they have seen "the whole thing" – that makes it so difficult to define the boundaries of a netprov. During a social media-based netprov, pieces of the fiction appear interwoven into each user's normal feed, coexisting with personal and public events of the day. Even though searching a netprov's hashtag presents to the user posts that are so tagged, on the way in and out of this search other posts appear and become part of the experience. This is a very fuzzy boundary. In line with a number of Avant Garde art theories of the past hundred years, each participant in a netprov experiences a materially different work with variations including things like the typographic settings of the participant's appliance (phone, pad, laptop, etc.), the participant's search habits, the participant's chosen feed, and what the platform's algorithm is currently serving up.

Modeled on the use pattern of Twitter itself, the netprov *#1WkNoTech One Week, No Tech* was perfect for partial reading, and short- or long- term participation because the cumulative aesthetic effect was built out of the jumble of individual posts and two- or three-message mini-narratives. *All Time High*, on the other hand, depended on a sense of sequential time as players progressed each week of the month-long project toward graduation. This sense of time was particularly acute during the weekly, hour-long, live events, The Talent Show, The Big Game, The Big Dance, etc. since these events come with their own sense of real-world timing and outcomes. The more consistently a netprov player was able to participate in *All Time High* for the entire month, the richer the experience became. Because of the importance of the narrative sequence in *All Time High*, the organization of the archive of such a netprov is crucial. Mark and I have kept private archives of these projects, but they are raw and in reverse chronological order, which we are used to in social media, but which is the opposite of how most written chronologies operate. It's a good thing we have kept these, too, since many of the character accounts netprov players created have been eliminated by Twitter as of this writing, and their wonderful utterances erased.

As a society we rely almost entirely on the large social media platform's own archiving systems. In doing this, we exhibit an enormous naiveté. *Of course, Facebook will keep all my posts and pictures (which, remember, they co-own) and give me access to them,* we blithely assume. *Of course, Twitter will keep everyone's tweets and continue make them searchable by hashtag.* None of this is certain. The 2014 Tumblr-based netprov *Mem-Eraze* invited players with this concept: "*Mem-Eraze* is an online support group for those who lost their online social scrapbooks in the Mem-or-Eaze Inc. server fire and bankruptcy" (2014).

The page went on to explain how to join the creative play:

> Recovering from the shock of losing all their family photos and despairing of being able to recreate past travels and past reunions, members of the Mem-Eraze support group begin to reconstruct family history using images of everyday objects — for one family the salt and pepper shakers become grandpa and grandma, soup spoons are the kids, teaspoons are the grandkids. They lay out in the Summer of 2014 on the napkin-beach by the tablecloth-sea. The family stories emerge vividly, nonetheless. In part a cautionary tale about entrusting our family archives and family narratives to online corporations — when will the first catastrophic loss of data occur? — this netprov harnesses the expressive power of the highly developed contemporary art form of snapshot-and-caption. (2014)

Mem-Eraze served as a caution about the potential transience of corporations' online archives. But it also was intended to encourage netprov participants to become aware of, and play with, the graphic design form and content of a social media archive.

Encouragements for netprov archiving (and social media in general)

My first encouragement concerns graphic design. Most social media platforms are not very well designed. At all. I'm not just talking about the clunky esthetics, crummy colors and bland typography. I'm pointing out that they don't take advantage of the best thinking about how to organize and make visible large data sets. Yes, you could uphold the antiquated view that Twitter posts should continue to look like classified advertisements in an old newspaper and that Facebook posts should continue to look like an early blog imitating a traditional print magazine, with black and blue ink on the white paper of the screen.

Wouldn't you rather see a sweeping overview of recent activity among your friends since you last logged on – literally what you have missed? Wouldn't you rather be able to zoom in a bit closer and see where your favorite writers, or perhaps your current favorite hashtag, feature in the flow? Wouldn't you rather be able to see an overview timeline showing how much action there has been in a netprov hashtag for the last 24 hours and be able to zoom in on the most active part? Wouldn't you rather be able to zoom through your whole feed with the netprov posts brought forward brightly and the rest backgrounded and still readable? Wouldn't you rather be able to request a replay of the netprov and have the hashtagged texts appear in your feed with the same exact timing they did the first time around? This handful of ideas is a mere suggestion of how good, how useful, how pleasurable social media archives could be if they were well designed. Good archives could give users a sense of what the netprov was like in real time.

Don't take as a given the way a social media platform designs its flow. Re-conceive it. Re-design it, when you archive social media work. Go for inspiration to astonishing studios of information graphic design such as the New York Times' interdisciplinary visual storytelling team. By aiming to be faithful to the work, you'll be forced to make intelligent decisions about what belongs in the archive and what doesn't. From this process will come the best answer about the size and borders of a netprov.

My second encouragement refers back to the question of how long should a netprov be? The netprovs Mark and I do begin months before the launch as concepts that percolate in conversations and planning sessions. Then a couple of weeks before the public launch announcement we usually begin experiments – rehearsals if you will – on a private platform among the featured players, which is what we call the folks who regularly play with us. The best of these rehearsal texts is re-copied into the live work once public play has begun. Even the public launch has come to consist of a soft launch, where featured players begin to play in public, and then the hard launch, the announced start date. In other words: the beginning of a netprov is porous, organic, cloudlike. Then, we usually announce an end date for the netprov. This is based on how much time Mark and I and other netrunners can afford to play the netprov full-on, while we're working our regular jobs. This is usually a couple of weeks.

But then the netprov itself takes over. One of our favorite phenomena over the years is that netprov players often ignore our proposed end date. They fall in love with their characters and don't want to bid them farewell. They are still playing, having fun, creating new narratives. And who are we to stop them? TV showrunners declare and end to production when they stop paying people. Since netprovs so far are voluntary and unpaid, netrunners don't really have the right to shut them down. Other players can repeat netprov concepts on their own and play at will. Despite all these cautions and caveats, I will offer one person's answer to the question.

To my eye, these days, the optimum run time for a netprov is about one month, with a soft launch and ramp-up period, a couple of weeks of really concentrated creativity in the middle and a gradual tapering off. And if your players fall in love with their characters – let the characters live on!

My third encouragement is also a personal request for pointers to make netprov better. Other perspectives, please! In interdisciplinary fields like netprov it can be hard to find the others who are out there doing inspiring projects. I'd love to read what theater and performance studies folks and game studies folks do that is netprov-like and what they would write about the experience of netprov in real time. I fear my literary background makes me focus too much on the archive "text." Let's join forces! I'd love to see an interdisciplinary team create an anthology of netprovs from these different traditions. No, seriously, contact me through the publisher. I would love to learn what such a scholarship could offer to help make better netprovs. I would love to read what information designers would think about how to design a living archive of social media, more vivid and accurate than the sterile tire tracks we have now.

Works Cited

Marino, Mark and Rob Wittig. "#1WkNoTech One Week, No Tech: A Netprov." Meanwhile netprov studio. November, 2016. http://meanwhilenetprov.com/index.php/project/one-week-no-tech/.

_____. "#fixurl8tionship Fix Your Relationship (For the Camera): A Netprov." Meanwhile netprov studio. April, 2018. http://meanwhilenetprov.com/index.php/project/#fixurl8tionship.

_____. "All-Time High: A Netprov." Meanwhile netprov studio. July, 2015. http://meanwhilenetprov.com/ath15/.

_____. "Mem-Eraze: A Netprov." Meanwhile netprov studio. May, 2014. http://meanwhilenetprov.com/index.php/project/mem-eraze/.

_____. "Press Release for All-Time High Participatory Netprov." Meanwhile netprov studio. July, 2015. http://meanwhilenetprov.com/ath15/index.php/about/press/.

Part III.
Interactivity in Action: Current Case Studies

Chapter 7

Text Sonification and the Literacoustics of Language-to-MIDI

Kedrick James, Esteban Morales, Rachel Horst, and Effiam Yung

University of British Columbia, Vancouver

Abstract: This chapter explores the sonification of linguistic data as a digital arts-based educational research practice. The transmediation of language into digital sound forms as a way of composing collaborative interpretations of texts will be instrumentalized using voice recognition and MIDI-based algorithms so that spoken words can be simultaneously, synthetically, musically performed. Particular focus in this chapter will be given to the manner in which textual characteristics are construed as MIDI code. The acoustic mapping of linguistic data, stripped of its semantic load, is carefully delineated. And yet this work celebrates error and serendipitous accident in the synaesthetic transmediational process, seeing these irreducible glitches as topoi where creative collaboration between human and other-than-human entities become visible, audible, available to improvisation, and viable as instantiation of a new aesthetic direction in digital literacy education.

Keywords: Literacoustics, Sonification, Lexicogrammar, Sound Objects, MIDI

***

1. Introduction: Literacoustics and Sounding Text

The differences between sensory perception of sound and of light reach deep into human epistemologies and behaviours; bias toward one mode of perception over the other — literacy versus orality — is a cultural determinant. This dichotomy has been hotly debated throughout the electronic media era with its proliferation of digital literacies and the various automations of our most sacred form of communication in a visually biased culture— writing. Of the many modernist scholars central to these debates[1], most tend towards a framing of what Ong termed the "psychoacoustics" of language as distinct from the visual perception and interpretation of written systems. Ong predicted a "secondary orality" (which has indeed come to pass) and the resurgence of spoken language which would become interpolated with written forms in dominant media for public messaging (radio, tv, social media, etc.).

Within literate cultures, other scholars such as Halliday sought to define "another level of organization in language to which both the sound system and the writing system are related" (17). Furthermore, Halliday writes, "[t]he sound system and the writing system are the two modes of expression by which the lexicogrammar of a language is represented…" (7).

In the Digital Literacy Centre (DLC) at the University of British Columbia, we are attuning ourselves to digital voices of texts, "thoroughly transmogrified" (Illich 36-37) by algorithmic agents, to grasp the educational implications of datafication of literacy under conditions of near-ubiquitous automation of all human communications. As Shaw axiomizes the data-centric worldview: "data speaks and the world responds accordingly" (2). To this end, we are developing Singling, a platform that allows us to create and manipulate auditory representations of the lexicogrammatical properties of a text, i.e. the implicit connection between form and meaning, or syntax and semantics, of language in use. Our conceptualization of the cultivated skill necessary to read these auditory representations is termed *literacoustics*— a literacy that *others* a strictly literate decoding of a text. This literacy practice allows us to apprehend and understand texts at the level of their lexicogrammatical structure, which becomes manifest with the help of algorithms. The sounds we read are not those of a voice pronouncing words, but rather an other-than-

---

[1] The following modernist scholars (among many) provided rich debates around psychological and ontological implications of the alphabetization of the mind:

Goody, James. *The Interface between the Written and the Oral.* Cambridge University Press, 1987.

Havelock, Eric A. *The Muse Learns to Write: Reflections on Orality and Literacy from Antiquity to the Present,* Yale University Press,1986.

Innis, Harold Adams. *The Bias of Communication.* 1951. Intro. Marshall McLuhan. Toronto: University of Toronto Press, 1964.

Jaynes, Julian. *The origin of consciousness in the breakdown of the bicameral mind.* Penguin, 1976.

McLuhan, Marshall. *The Gutenberg Galaxy.* Random House, 1969.

human sounding of "the abstract level of coding" that is "[t]he heart of language" (Halliday 194). As Wittgenstein wrote so evocatively in *Prototractatus,* "there are, indeed, things that cannot be put into words. They make themselves manifest. They are what is mystical'' (6522). With Singling, we are attempting to read these things about words that cannot be put into words— the ineffable communicative energy of language. We approach this work as linguists and musicians, composers of sound and written systems, but moreover as arts-based researchers keenly interested in new ways to apprehend, analyze, and present/perform textual data.

While Singling is a work-in-progress with many features still to come, it already sings to the rich potentiality of sonification for the representation of textual data and literacoustics as a way of reading those representations. While literacoustics is related to aurality, it is a natively digital concept—computers must exist in order to run the algorithms that transform the textual data with precise repeatability. A complex webwork of algorithms performs the heavy lifting of transmediation as we shift from graphemic text to the sounded structures of string data. In this sense, literacoustics is a new approach to literacy and worthy of study because it affords a fresh way of apprehending meaning in texts— meaning that is to some extent occluded by more traditional text-based sense-making. Singling plays language patterns of form and content that become available through algorithmic transmediation. This kind of data listening has parallels in geology and the audification of earthquake data or tidal analysis (Romans), network monitoring or intrusion analysis (Debashi and Vickers), and communications studies looking at the rate, for example, of posts, comments, and updates on Wikipedia (Hashemi) to name just a few.

The desire to harness digital technologies towards *othering* traditional reading practices and seeking new meaning and insight underlies Moretti's distant reading, a conceptualization of reading that posits distance as "a condition of knowledge" allowing one "to focus on units that are much smaller or much larger than the text: devices, themes, tropes—or genres and systems" (57). Distant reading stands in opposition to the close reading methods of literary analysis and seeks to make intelligible entire canons of literature at the expense of many features of the individual text. Moretti writes that "if, between the very small and the very large, the text itself disappears, well, it is one of those cases when one can justifiably say, Less is more. If we want to understand the system in its entirety, we must accept losing something" (57). In this, he champions what Ong and his contemporaries often bemoaned, the loss of an embodied gradient of knowledge that visualizing language quickly suppressed. Visualizations based upon distant reading algorithms that quantify text features such as word frequency have evolved well past their faddish word cloud stage to serve in more complex folksonomies (Kumbhar) used for content sharing networks such as Reddit and (meta)tagging of large databases with consequential flattening of information (Juliano and Srinivasan). Others have broadened this type of analysis to notions of betweenness and closeness of terms in the linear composition of language, rending live data feeds expressed in pliable and filterable rhizomatic webs. These tools capture texts and tame them, driving them toward a singularity, an immanent totalizing instant; part of their interest value is that they are completely stripped of affect, yet can claim to measure it, to view *the system in its entirety.* But what of the ebb and flow, shifts and changes, both subtle and obvious, that are only apparent over time? With Singling, we do not accept the losses of distant reading in terms of a text's temporality and individuality. Rather than seeking clarity through physical distance from the text, Singling enables deep immersion in its ecology. The literacoustic reading of the text does not seek omniscience, having a god's eye view into the vastness of a canon; it individuates each text's unique lexicogrammatical character; and in this way, literacoustics attends to the individuality of a text through a process of othering ourselves from its literality.

2. Background: from The Patch to Singling

Prior to the creation of Singling, the DLC developed The Patch (James, Horst, Takeda, and Morales), a creative w/reading (Torres) workshop that guides participants through a series of processual engagements and collaborations with a given textual data set. The Patch is a collaborative human-computer data performance matrix used to understand literary texts through the generation of multiple poetic singularities–instances which encapsulate an othered reading practice that emphasizes visual and embodied, haptic engagements interspersed with and distorted by regenerative or transmediational algorithms. The last state of transformation of the writing sequence in the workshop is sonification and amplification, when the collaborative poems are transformed into MIDI code and sounded with a MIDI instrument upon speakers. Initially, this transformation was enacted using Textsound (Kohll) a program on github.com, which converts the roman alphabet to Musical Instrument Digital Interface (MIDI) code based on the overtone series. This little gem of a program was built on JFugue, an open-source library used for programming music in Java. We created a small, glitchy interface which allowed us to create stop words and set the range parameters for duration, tempo, octave and frequency range. It used the default MIDI grand piano #001 instrument. We would take the poetic recasting of the text, open it with our user interface for TextSound, and play the MIDI file with the default Windows Media Player. For an example of a Patch workshop collaborative output, see https://youtu.be/EO202QiYEYA.

Using the overtone series according to machine, reading/translating one letter at a time into the sonic gestalt of a grand piano, produces a composition that bears some resemblance to microtonal compositions by 20th Century avant-garde Western art music composers–think Cage, Boulez, Feldman, Stockhausen, at times slow and sweet as "Gnossienne" by Satie or frenzied and moody as "Evryali" by Xenakis. It is an abstract, floating, often melodious soundscape, and has room to breathe because spaces trigger "note off" messages that let notes linger. Previous experiments with this software had interesting results when outputs were played to communicate with hummingbirds (James, "The piano plays a poem"). While we enjoyed the aesthetic experience of these transmediations, we felt the need to construct softwares that would allow us more range and freedom for textual analysis and expression than TextSound alone afforded us. Singling began to materialize in the naming and categorizing of our specific needs, the various uses we imagined for such a program, and the deconstruction of our initial impulses (that generated new impulses and desires) so that we might develop and instrumentalize them in code.

3. The sound of a text: Unfreezing the world

The word sonification — or sound with meaning, a sound-text — can indicate an auditory signal such as an alarm bell announcing danger, the beeping of a truck backing up, or a crosswalk indicating that it is safe to cross. Increasingly, however, sound is being harnessed to communicate far more complex forms of messages and data. Traditional forms of data representation are not robust enough to contend with the complex phenomena that are embedded in the vast amounts of readily available data nowadays (Ferguson and Brewster); and, indeed, increasingly powerful media technology combined with progressively larger datasets made it necessary for new modes of interacting with information to arise (Kramer et al.). It was in this scenario where sonification, as we know it today, was born. Kramer et al. define sonification as "the use of nonspeech audio to convey information. More specifically, sonification is the transformation of data relations into perceived relations in an acoustic signal for the purposes of facilitating communication or interpretation" (4). Sonification is a sub-type of auditory display, a term that covers any form of information representation via sound. Sonification has four particular properties that make it an especially valuable approach to data representation: 1) it reflects objective properties and/or relations in the input data, 2) its transformation is systematic, 3) its results are reproducible, and 4) it can be used with different data or in repetition with the same data (Hermann; Walker and Nees). Furthermore, sonification refers to both the *technique* as well as the *process* of representing information with non-speech audio; in the words of Hermann, sonification is analogous to an "algorithm that is at work between the data, the user and the resulting sound" (3).

Mapping metaphors permeate the literature of sonification and data display in general. Navigable pathways are necessary to make meaning of the enormous computationally derived data sets that are now at our disposal. Data requires an interpretive map in order to navigate the terrain. Any single map, however, is not a definitive representation of the data; each mapping is but one instantiation of innumerable possibilities. Drucker, like many in the digital humanities, has been critical of the "representational force" of graphical mappings that tend to fix the interpretative quality of data into the bounded geography of "divisions, coordinate lines, scale figures, circles, rectangles, curves, bars"– as though these are objective "pictures" of the data (5). Drucker argues for a more "humanistic, interpretative, and constructivist graphical expression..." as opposed to "realist statistical graphics," in which "the curves, bars, columns, percentage values would not always be represented as discrete bounded entities, but as conditional expressions of interpretative parameters–a kind of visual fuzzy logic or graphical complexity" (Drucker 6). Thus sonification surfaces as a possibility to complement or even replace a graphical mapping practice that tends to "freeze the world" (Palmer and Jones 223). In this sense, Lodha, Wilson and Sheehan outline some of the aspects in which sonification can outperform visualization:

> Sonification can often provide information about data that cannot be seen using visualization. Sound can enhance a graphical presentation by providing information about features of the data that may be hidden or occluded. Sound can also help the user to distinguish the size relationships between objects that may be difficult to determine visually because of projection distortion. In multivariate mappings, where visual clutter can destroy the usefulness of the display, the addition of sound can provide relief for the overloaded visual channel by allowing some variables to be presented aurally. Also, sound can provide redundancy of representation and allow data validation (3).

Sonification then proves to be a comprehensive tool to convey meaning, showing that it can be particularly valuable when displaying temporal and rhythmic elements (Neuhoff) when there are hidden events or objects in the data (Minghim and Forrest), or with uncertainty in the data (Ballatore, Gordon and Boone).

4. Developing Singling's Database

In order for a software program to transmediate text to sound, it must be able to parse the input data —which requires a database in order to correlate instances in the data with a range of known tokens. Accordingly, we started to build our database from the smallest data points —letters, symbols, punctuations— until we reached more complex categorizations of language. Our mapping of features and categories of language from the formally granular to the broader conceptual categories have been part of an evolving discussion. Indeed, this is a lexigrammatical approach to language ontology that seeks to bring unexpected insights to our understanding of linguistic knowing and expressing. Singling has the following collection of databases that we gathered, modified, and curated.

5. Letters, Symbols and Punctuation

Our initial work with TextSound (Kohll) was based upon letter frequency and the sonification of individual alphabetic characters of a text mapped to the overtone series (see Figure 2, below). Letter frequency is a feature of text often studied by cryptographers and is useful in modern data-compression techniques such as Huffman coding. Letter frequency data can only be gleaned by analysis of large amounts of representative texts, a task well suited to modern computing, and has been shown to be a kind of DNA test of the writer and subject of the text. "Different authors have habits which can be reflected in their use of letters. Hemingway's writing style, for example, is visibly different from Faulkner's" (World Heritage Encyclopedia).

Accordingly, we have included three different alphabetic frequencies derived from three different methodologies. Mayzner, in his widely cited publication *Tables of Single-letter and Digram Frequency Counts for Various Word-length and Letter-position Combinations,* culled a corpus of 20,000 English words for his letter and word frequency tables. In 2012, the 85-year-old Maynzer reached out to Norvig at Google to use "the computing power that is now available to significantly expand and produce such tables as [he] constructed some 50 years ago" (Norvig). This frequency listing is 37 million times Mayzner's 20,000-mention collection in size: ETAOINSRHLDCUMFPGWYBVKXJQZ. We also draw upon the widely referred to standard order frequency used by typesetters ("Letter frequency"): ETAOINSHRDLUCMFWYPVBGKQJXZ. And lastly, we include a letter frequency based upon the Oxford dictionary ("Oxford Languages: The Home of Language Data"): EARIOTNSLCUDPMHGBFYWKVXZJQ. In addition to individual alphabetic characters, punctuation and the most commonly used symbols were added as separate columns so that Singling would be able to differentiate between them, thus creating a rich array of graphemes that can be identified and sonified.

6. WordNet

There are many use cases for examining and experiencing language at the level of the grapheme, but it does not emancipate the force of utterance from the textual field. When we discovered the lexical ontology of WordNet, a linguistic database created by Princeton University's Cognitive Science Laboratory (Princeton University), the scope of our textual territory radically shifted. WordNet, a vast, decades-long natural language processing (NLP) project, is a semantic network that interlinks words employing lexical and conceptual relations (Fellbaum). Like many research groups before us, "the availability of the WordNet database was an important starting point... [because] WordNet covers the entire English lexicon and provides an extraordinarily large amount of conceptual distinctions. As well, it is particularly useful from a computational point of view because it was developed for easy access and navigation through its hierarchies" (Strapparava and Valitutti 1083). However, many of the features of WordNet's rich ontology were not relevant to our purposes. Singling has retained two key categorical attributes of the corpus: word class (nouns, verbs, adjectives, and adverbs) and the 45 LexName categories, which are a word's syntactic and logical grouping. For example, nouns are organized into 26 different LexName categories including body nouns, animal nouns, and person nouns.

Word meaning is an ongoing problem for computational lexicographers because "[w]ord meaning is in principle infinitely variable and context-sensitive. It does not divide up easily into distinct or discrete sub-meanings. Lexicographers frequently discover in corpora loose and overlapping word meanings, and standard or conventional meanings extended, modulated, and exploited in a bewildering variety of ways" (Edmonds and Eneko). While we have retained WordNet's LexName identifiers, we are using them in a very different way, and for this reason, have renamed them Lexicogrammatical Categories (LGC). Singling is indebted to Wordnet for the ability to distinguish the grammatical function of each word in the major word classes, the lexicogrammatical categories each of those words belongs to, as well as the breadth of a word's polysemy, or the quantity of categories a word may belong to.

7. Sentiment Analysis and Emojis

After the reorganization of the WordNet databases to best fit our work on Singling, we began exploring how sentiment analysis could be integrated into the program. To do so, we added SentiWordNet (Baccianella, Esuli and Sebastian), which is compatible with WordNet and contains an automatic annotation of positive and negative dimensions of many adjectives. The annotations in this database, assigned with a value between absolute 0 and 1 to each dimension, are not limited to a single value per word but assess how the same term might have different opinion-related properties. For example, SentiWordNet considers the word "Common" to be somewhat positive (0.25) and somewhat negative (0.25) at the same time, showing how it can consider different (albeit, limited) contextual and semantic features of the words. Singling combines the values of both connotations and their sum (mathematical difference) results in a major chord if overall positive, and a minor chord if overall negative. Each tenth (0.1-0.2, etc.) of a degree (in the shift of mood) translates to a different type of major or minor scale. Furthermore, we added emojis to the collection of Singling's databases. For this, we used Novak and colleagues' database, which includes the most used emojis based on an analysis of 1.6 million tweets. Additionally, as the database weights emojis based on sentiment polarity (3), we reorganized the emoji database following SentiWordNet's system.

8. Function Words

Other functional word classes have been added to our database collection, including articles, modals, pronouns, conjunctions, and prepositions. To achieve this, we used publicly available databases, such as the list of prepositions in Wiktionary ("English prepositions") that contains 650 words and phrases. Out of this process, we eliminated all multi-word phrases since they are currently incompatible with Singling's parser which uses spaces to identify individual word tokens. Furthermore, and looking to add literacoustic nuance to Singling prepositional analysis[2], we open-coded the preposition database, resulting in three new categories: 48) agential prepositions that denote a doing or making or resisting capacity in relation to the self, e.g. *by*; 49) spatial prepositions that orientate subjects and/or objects in space, e.g. *beyond*; 50) temporal prepositions that orient relations of the subjects and/or objects in time, e.g. *after*. Finally, and looking to further explore the semantic and affective dimensions of the textual data fed to Singling, following James ("Toward a Theory of Choice through Visuals"), we added sentiment data to the modals utilizing a simple pattern where 1 has a value of most necessity (for example, *Must* is 1.0 positive and *Must Not* 1.0 negative), and values closer to 0 represent

[2] The motivation for this added classification of prepositions derives from their significant metaphoric capacity regarding orientation (thus, embodiment) embedded in common English, and that when combined with many common verbs, exhibit viscerality in colloquial expression, (e.g. "throw up" as opposed to vomit). See Lakhoff and Johnson, in *Metaphors We Live By* and later *Philosophy of the Flesh*, on the embodied metaphoric properties of prepositions.

possibility (*Would* is 0.1 positive and *Would Not*, 0.1 negative). Table 1 shows the final list of linguistic components and LGCs used in Singling.

Table 7.1. LGCs in the database of Singling, modified from "LexNames". Fields highlighted in blue means that these LGCs were modified by the authors, and fields highlighted in red were added.

LGCs	Category	Details
00, 01, 44	Adjectives	[00] Adjective all, [01] pertainmyms, [44] participles
02	Adverbs	[02] Adverbs all
03 - 28	Nouns	[03] Noun tops, [04] acts, [05] animals, [06] artifacts, [07] attribute, [08] body, [09] cognition, [10] communication, [11] events, [12] feeling, [13] food, [14] group, [15] location, [16] motive, [17] object, [18] person, [19] phenomenon, [20] plant, [21] possession, [22] process, [23] quantity, [24] relation, [25] shape, [26] state, [27] substance, [28] time
29 - 43	Verbs	[29] Verbs of body, [30] change, [31] cognition, [32] communication, [33] competition, [34] consumption, [35] contact, [36] creation, [37] emotion, [38] motion, [39] perception, [40] possession, [41] social, [42] stative, [43] weather
45	Modals	[45] Modals all
46	Symbols and punctuation	[46] Symbols and punctuation all
47	Emojis	[47] Emojis all
48 - 50	Prepositions	[48] Agential prepositions, [49] relational, [50] time
51	Pronouns	[51] Pronouns all
52	Articles	[52] Articles all
53	Conjunctions	[53] Conjunctions all

9. Sonified texts: transformation processes in Singling

Along with the work of integrating new linguistic features into the Singling database, we explored different sonification processes to transform the identified textual features into sound. Users are able to control various base settings on the main screen of the interface (see Figure 1). Transformations (available on the right side of the interface) allow for more precise and customisable control of changes to the sonic output. The user-selected assemblage of base settings and transformations create the pattern by which Singling transmediates the textual data (entered in the center section of the interface) into MIDI code. Once the algorithmic process has been designed, the sonic output is determined and replicable. The lack of random interference in transmediation, combined with this repeatability of output, ensures that Singling can be operationalized as a data sonification instrument (Hermann; Walker and Nees). Singling can sonify textual input in real time as the text is being processed (stream mode On), or processed as an entire file that can be paused, allowing the user to adjust the base settings during playback.

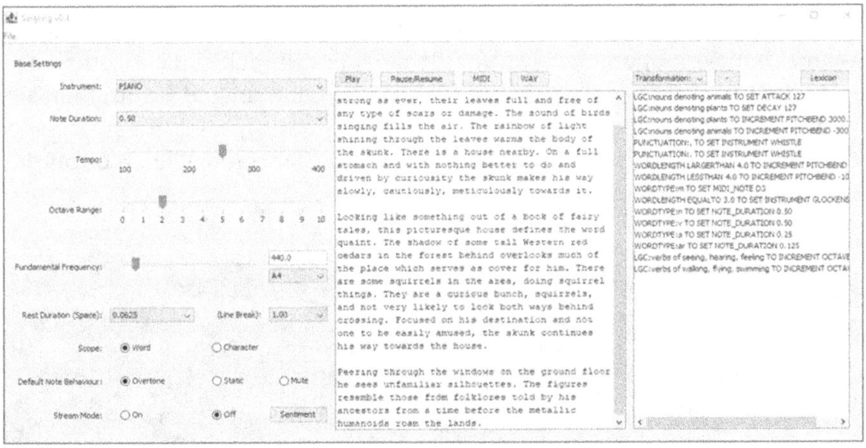

Figure 7.1. Graphical user interface of Singling version 0.4.

10. Base Settings

Base settings relate to generic features of musical sound. The user selects a digitally-emulated instrument then specifies the duration of each note played (from whole notes to semi-hemi-demi-semi-quavers), the speed of which is determined by the tempo setting. Some instrument types cannot, however, play extremely rapid notes, some cannot sustain notes, and so on, because of the physical constraints on sound production for that instrument type in the real world. One can also choose the duration of the rests (note-off) triggered by spaces between words, and by line-breaks. Users can set the fundamental frequency (lowest note in a chord, first note in a scale), and the octave range within which different notes are sounded. Similarly, constraints on sound production by real-world instruments also affect the MIDI output, for example, a violin's strings are not long enough to conduct slow (i.e. low note) vibrations for which violas, cellos, or double bass viols were invented.

Frequency (wavelength) is the basic unit of Singling because it has a value that can be easily encoded to output a note programmatically (Wolfe). Using jFugue, we convert the frequency of each word or character to a combination of MIDI note and microtone; these calculations are normalized to fit within a target octave range defined by the user. As previously discussed, the scope of text sonification by Singling works either with individual letters according to preset frequency values, or with words according to the LGC values. Users also set the default behaviour of each note. In this case, users can choose from three default behaviours: 1) Overtones: Default mode of operation where notes (pitch) are determined by the ordering of alphabetic frequencies or by the numbering of LGCs in relation to the overtone series (see Figure 2) 2) Static: Every note will sound the same, determined by base settings, unless otherwise instructed by sound transformation(s); 3) Mute: Notes are muted unless otherwise instructed by sound transformation(s). These predetermined features enable users to interact with textual data in different ways, adjusting the process to different contexts of uses of Singling. Lastly, the user can toggle on or off the use of sentiment values.

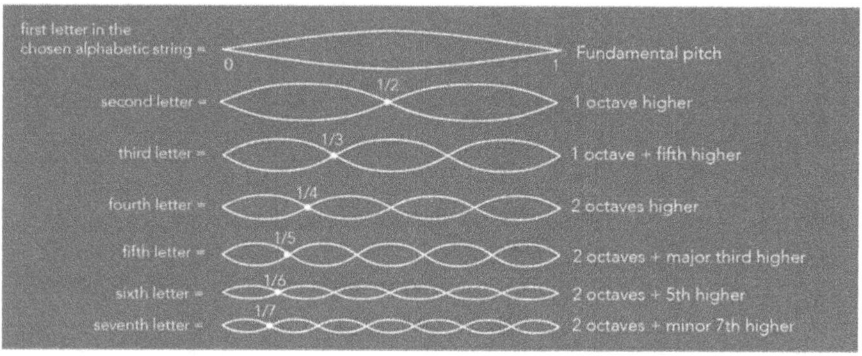

Figure 7.2. Alphabetic string as it corresponds to the overtone series represented as wavelength. Note that overtone series also applies to words according to the LGCs.

11. Transformations

Transformations are the core of Singling's potential as literacoustic software, giving the user the ability to specify which lexicogrammatical feature in the databases affects which MIDI parameters of the base settings and how, indeed how much, that change is instantiated (see Table 2). Each transformation opens up a dialogue box (see Figure 3) offering an initial choice between Set or Increment. A Set transformation will apply a constant MIDI value (such as a specific note, note length or instrument) each time the change is triggered by instances of the specified token in the source text. Increment transformations apply a chosen increase or decrease in the MIDI value each time the change is triggered (such as an increase in volume, or a decrease in tempo). On a mundane level, one can listen for misspelled words by triggering a sound only when a word is not contained in the database. For more complex textual analysis, one might listen for cohesion in a text by identifying adverbial conjunctions and trigger an octave change based on commonality (where "And" would trigger a note transposed by one octave, "Nevertheless" by three, or "Notwithstanding" by four octaves, and so on).

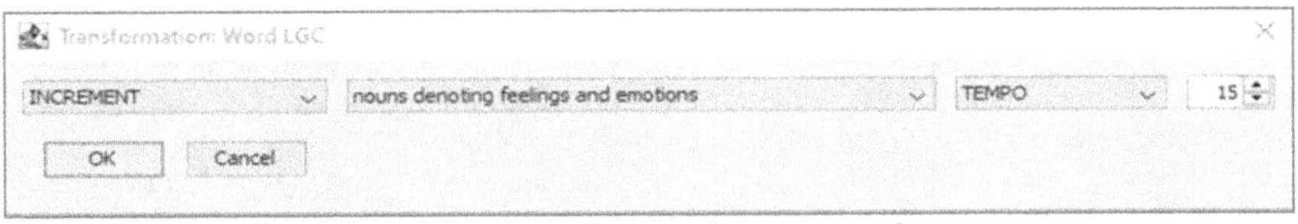

Figure 7.3. Transformation dialog box where each instance of LGC 12 increments tempo by 15 BPM.

Instances where words fall into multiple syntactic categories or LGCs result in additional channels or *voices* (as referred to by JFugue) sonifying each of the LGCs simultaneously with a maximum of 15 separate notes related to a single word in the database. Transformations in Singling are cumulative, hence it goes beyond the scope of this paper to address the colossal number of possible sonfications of any given source text given an unlimited number of specified transformations.

Table 7.2. Transformational features and their parameters in Singling.

Text feature types	**Comparison operators**	**Text feature value**	**MIDI parameters**
WORD_CATEGORY	EQUAL_TO	noun (n), verb (v), adj (a), adv (r), modal (m), symbol (s), prepositions (p), pronouns (pn), articles (ar), conjunctions (c)	TEMPO, NOTE_DURATION, OCTAVE, INSTRUMENT, VOLUME, MIDI_NOTE, ATTACK, DECAY, PITCHBEND, PAN
WORD_LENGTH	EQUAL_TO, GREATER_THAN, LESS_THAN	any integer	TEMPO, NOTE_DURATION, OCTAVE, INSTRUMENT, VOLUME, MIDI_NOTE, ATTACK, DECAY, PITCHBEND, PAN

LEXICOGRAMMATICAL_ CLASS (LGC).	EQUAL_TO	1-53	TEMPO, NOTE_DURATION, OCTAVE, INSTRUMENT, VOLUME, MIDI_NOTE, ATTACK, DECAY, PITCHBEND, PAN
SYMBOL	EQUAL_TO	list of symbols from database	TEMPO, NOTE_DURATION, OCTAVE, INSTRUMENT, VOLUME, MIDI_NOTE, ATTACK, DECAY, PITCHBEND, PAN
PUNCTUATION	EQUAL_TO	list of symbols from database	TEMPO, NOTE_DURATION, OCTAVE, INSTRUMENT, VOLUME, MIDI_NOTE, ATTACK, DECAY, PITCHBEND, PAN
CHARACTER	EQUAL_TO	consonants and vowels	TEMPO, NOTE_DURATION, OCTAVE, INSTRUMENT, VOLUME, MIDI_NOTE, ATTACK, DECAY, PITCHBEND, PAN

11. Saving the assemblage

When saving a Singling analysis, there are few options. One simple and lightweight approach is to save a java properties file, which preserves the configured input text, base settings, and transformations and can be reopened in Singling. Another method is to export an audio file. The third option is to export a MIDI file. When opening a Singling MIDI file in a MIDI sequencer software such as Aria Maestosa, the composition will automatically open with as many individual tracks as the number of notes in the most complex chord of the composition (this number correlates to the textual word with the highest number of lexicogrammatical categories and has an upper limit of 15, which is a MIDI constraint). Instruments are likely to be different for each MIDI track, and one can modify parameters in the sequencer, in an embedded software instrument (e.g. TAL Noisemaker), and in many cases, in the connected MIDI instrument itself. This hints at the great range of possibilities for personalizing the listening experience and honing literacoustic sensibilities.

12. Working with and against MIDI Parameters

The transmediation of language into music is more than just translation, although both have formalized writing systems. Music is built into MIDI, not just in name, but in terms of its basic functionality. The question of automating musical notation to control digital instruments presents some interesting challenges because it formalizes sound according to conventional musical traditions and scales. Even when visually represented, a MIDI sequence looks (and functions) much like a player piano scroll in a digital medium. The traditional piano, quintessential instrument of classical European musical tradition, is already mechanized because it removes the hands from the strings, so it already has a user interface (the keyboard): push a key, and an internal hammer strikes a string with an equivalent velocity, while pedals sustain or dampen the sound. When the piano is tuned, each note is a semitone apart from the note below and above as defined by European concert scale, centered on middle A at a frequency of 440hz (cycles per second). There's no bending of the pitch with embouchure and breath, or stretching a string with fingers, or applying extra pressure with a bow. The piano has mechanical exactitude, which is dramatically different from playing the shakuhachi, for example. There is a kind of correctness associated with it that bears a certain relationship to correct grammatical usage of language. This is not to say that piano doesn't invite invention, just as works of literature frequently break with grammatical and orthographic rules. Indeed, much of 20th-century classical music was inspired by Cage's idea to "prepare" his piano by glitching the mechanism, putting objects in, on, and around the strings, producing more percussive sounds, causing rattling or dull thudding, and creating compositions that aligned more with Southeast Asian gamelan than a baroque piano concerto. However, with the notion of music comes expectations on the part of the listener, expectations of conformity to the norms of genre that can seriously prejudice the listener.

MIDI was created in Japan in 1983, the miraculous banding together of many competing music hardware and software companies to develop a code that could be sent from a controller to any compatible device (The MIDI Association). If music is not what you intend, in other words, if you want to sonify but not compose, you must in essence work against standard MIDI conventions to get it to do things it wasn't intended to do. It should be noted that MIDI code contains no audio data. It sends two channels of data to MIDI-compatible instruments, one with specific note information capable of being distributed to 16 different voices, one with system exclusive messages and MIDI Time Code which synchronizes playback on devices. This adds a user-controlled flexibility in terms of how the code is converted into sound. The same MIDI track can be played back using different instruments, even triggering audio samples instead of instrumental sounds.

In our initial stages of conceiving the transmediation process, we began working with a kind of 1:1, or literal translation. For example, a space between words could cause a brief (i.e. a 1/8 note) rest, a comma could be a ¼ note rest, a period, a ½ note rest, a paragraph break could equal a whole note rest. A capital letter could increase volume (velocity of the attack) just as, when beginning a sentence or verbal utterance we give extra velocity to the attack of the first letter/word. A word with multiple meanings could play a chord instead of a single note. Vowels could cause a pitch bend, just as the voice adopts a rising and descending tone when saying WOW. A question mark could add sostenuto and a rising pitch bend. Word length could translate to the sustain-decay-release portion of the sound envelope related to note duration so that short words resulted in a faster, percussive tempo, longer words, a slower tempo, and so on. Table 3 shows all the MIDI transformations included in Singling.

Table 7.3. MIDI Transformation types and parameters in Singling.

Sound modifier	**Detail**	**Values**
TEMPO	Represents beats per minute (BPM) and indicates how quickly a song should be played. Transformations to tempo is always global and applies to all musical events that follow after.	[40, 220]
NOTE_DURATION	Indicates how long a note should be played. The values represent the note durations as follows: whole:1.00; half:0.50; quarter:0.25; eighth:0.125; sixteenth:0.0625; thirty-second:0.03125; sixty-fourth:0.015625; one-twenty-eighth:0.0078125.	"1.00", "0.50", "0.25", "0.125", "0.0625", "0.03125", "0.015625", “0.0078125”
INSTRUMENT	As specified by the MIDI code, there are a total of 128 different standard MIDI instruments. Singling currently utilizes 100 of these.	[0, 100]
OCTAVE	Represents a ceiling used to normalize notes to a reasonable range.	[0, 10]
VOLUME	MIDI controller event which specifies the volume.	[0, 16383]
MIDI_NOTE	Represents the alphanumeric note names (e.g. C4, D#5) which corresponds to a MIDI note number.	[0, 127]
ATTACK	Represents the attack velocity of the note, meaning how quickly the note reaches its full volume.	[0.127]

DECAY	Represents the decay velocity of the note, meaning how quickly the note decreases from its peak volume after it has played for its full duration.	[0, 127]
PITCHBEND	Used to make microtonal adjustments for notes.	[0, 16383]
PAN	Represents the location within the stereo field, where 64 is the center, 0 is the left and 127 is right.	[0, 127]

With the range of musical parameters that MIDI puts at our disposal, we can work with envelope (attack, sustain, decay, release), note and rest durations, tempo and time signature, key signature, octave and pitch range, velocity, pan, and amplitude, the instrument and timbre, the layering of notes into various scales and chords for harmonies, and so on. In the hard-coded base setting of Singling, we implemented a slider for base frequency. This allows users to have the fundamental frequency and subsequent overtone series not conform to Western musical scales but work in microtonal relation to it. The implementation of major and minor scales as a way to sonify sentiment values (in the case of adjectives) or necessity (in the case of modals) triggers different harmonic series in the production of chords (there are 8 major and 8 minor scales, each with subtle though affect-laden differences), with diminished (flattened) or augmented (sharpened) notes in the overtone series. Combined with a slider for the fundamental frequency of the chord, and the different tibral sonorities and envelopes of the preset instruments which can also be altered through lexicogrammtical transformations, Singling effectively escapes genre conformity or classification. This othering of the listening experience perks curiosity and resists the lulling, disattentive, background-sound aesthetic often used in live data feed sonifications.

13. The musical dilemma: on aesthetics

Sonification has a place in research as a useful method for the transformation of difficult to interpret quantitative data sets into sympathetic vibrations that can be immediately perceived and connected with in a human way. Like visualization, sonification raises questions around the use of musicality and aesthetics, where creative decisions, such as the choice of instrument or musical scale to include in the algorithm, could lead to very different interpretations or emotional experiences for the listener/data analyst.[3] While

[3] See Kaiser and Keller's work on the impact music can have on "the perception of normal and exaggerated expressions of happiness, contentedness, sadness and anger in … visual stimuli" (270)

these issues remain a topic of discussion in the sonification field (Walker and Ness), we acknowledge Barrass and Vickers stance that "thinking of aesthetics as being the framework for making decisions about artistic value and taste is unhelpful in this context [of data analysis] because it limits what we can do and even diverts our thinking, thereby distracting us from considering what aesthetics can be used for: the design of effective sonifications that promote sense-making, understanding, and pattern recognition" (159). Aesthetic choices, conditioned by the desires of the researcher, can filter the voice of the data itself through those preferences.

To further explore the difficulty of aesthetic choices in sonification, an example might prove to be useful. Amid the COVID-19 pandemic, a group of scientists from the Massachusetts Institute of Technology created a sonification of the virus' protein spikes, assigning notes in a musical scale to each amino acid of the coronavirus. The sonification was created to "help scientists find sites on the protein where antibodies or drugs might be able to bind—simply by searching for specific musical sequences that correspond to these sites" (Venugopal). The researchers decided to use the Japanese Koto as the main instrument (accompanied by chiming bells, the twanging strings, the lilting flutes) to play the score and show "soothing sounds that might bring some comfort in a time of trouble." Following Barrass and Vickers, this aesthetic decision to create a soothing song interferes with the goal of showing patterns embedded in the song's rhythm. Accordingly, these scientists might be better described as composing music (concerned with the aesthetic experience) rather than sonifying the data (where the emphasis is on the revelation of the phenomenon by listening to the sounds). In addition, it implicitly suggests a correlation between Japan (an inextricable feature of Koto's sonic gestalt) and the cause of a global epidemic.

As arts-based researchers, we have come across this quandary of aesthetics many times. Do we seek aesthetic satisfaction in our experimentations with data, or do we seek knowledge, or both? Does one borrow from the other? Are aesthetics and utility friends or foes? The very notion of arts-based research begs these questions. A central distinction might be made in terms of the motivation of the sonic display. "While numerous composers use mapping and other procedural techniques of one kind or another in their musical compositions, they are rarely interested in 'featuring' the mapping explicitly. Nor do they use mapping in order to simplify the working process or to improve production efficiency, but so as to craft the emergence of musical forms" (Worrall 17). While we agree with Barrass and Vickers that aesthetics and the purposeful shaping of music forms is not the goal of our work, nor should it guide our choices as we map the data —there is no avoiding that the resulting soundscape may be experienced *as* music. The very human, visceral, and aesthetic experience of *listening* is central to sonification as arts-based practice/praxis. In their work on arts-based research as a praxis for life, Visse, Hansen, and Leget paraphrase Gadamer who sees "art as experience and event… he is not interested in art's pleasure of beauty (a work's aesthetic properties) but in how art is a partner in conversation. This is about being phenomenologically involved in an artwork, resonating with it, receiving it, and being transformed by it" (9).

Literacoustics is thus a pragmatic phenomenological praxis. We encourage (and indeed exalt in) the complex assemblage of human perception that the listener/analyst brings to the event of listening. We acknowledge the transactional nature of engaging with the sonic event; indeed, literacoustics is a literacy, in which the listener hones and develops their analyses as a w/reading practice. We liken the literacoustic event to Rosenblatt's transactional theory of reading. She explains, "Every reading act is an event, a transaction involving a particular reader and a particular configuration of marks on a page, and occurring at a particular time in a particular context…. The 'meaning' does not reside ready-made in the text or in the reader, but happens during the transaction between reader and text" (4). She identifies two stances towards a text, the efferent stance in which "attention is centered predominantly on what is to be carried away or retained *after* the reading event" and the aesthetic stance, in which attention is focused "on what is being lived through *during* the reading event" (5). We encourage listeners to exploit the full range of the efferent/aesthetic continuum of attention in their listening, whether for pleasure or analyses or both.

The efferent/aesthetic w/reading of sound via the ear activates visceral emotional response deeply embedded in the human body. Whereas we are comfortable with the split-second juxtaposition of distinct images (up to a point), in general people find a rapid change in sonic timbres *physically* disturbing. Sound is orientating while vision is directing, and humans tend to be good at establishing the narratives of diverse images, whereas with sound we tend to hear cacophony unless heard as a soundscape accompanied by a visual logic. As literates, we are indebted to vision, indeed, *imagination* is the conjuring of images, sonation is something different, a felt-sense, very personal and visceral because each proprioceptive body resonates differently. We feel with sound in an encompassing sensibility of the immediate. We are sensitive to tiny timbral changes or pitch shifts in tones more so than to small distinctions in colour tones. Of course, all this is a matter of training the eyes or ears. So literacoustics is just that, training the ears to read sounds in order to analyze texts in sympathetic vibration with it, hearing sharp beats of rhetoric or cool tones of vowels, or sweeping glissandi of adjective clusters, hearing the sentiment, the mood, the voice, in our guts, our viscera.

> Listening ... is an aesthetic activity that challenges the philosophical tradition of the West, which, according to film theorist Christian Metz, is based on a hierarchy between the senses which positions sound in the attributable location, sublimated to the visual and its linguistic structure. In that position sound is left to describe and enhance but never to do and become.... When we start to listen as a critical motility this position becomes untenable. Listening emancipated from the expectation to enhance becomes something else. It produces, it invents, it generates. It demands that the heard be more than a ghost of the visual, a flimsy figment of the imagination, soon dispelled. However, instead of denying the ephemeral quality of its object, it is the preference for the assumed substance of visuality that needs to be reassessed by focusing on the ephemeral exactly. (Voegelin 13)

The sound of texts cannot help but be affected by the gestalt that accompanies the instrument that produces the sound: it is a sociocultural web that takes hold of our heritage and genetic predisposition. These perceptions are open to change; of course, we can grow to appreciate sounds that were previously thought unfamiliar and unpleasant simply through repetition. But it is very difficult to completely disregard the aesthetic dimension of sound, namely music, and its relationship to taste as a matter of social distinction (Bourdieu). Perception of auditory vibration is a whole-body experience that is merely refined by the tympanum. What to do with musical sense when listening to data that makes no attempt on the surface to be musical? One does not put on recordings of electroencephalographs to be entertained. Listening to the brain's electrical crackle to read the state of its health is only a clinical workplace activity. But amplify, distort, filter and perform it, and very quickly, we create music that a particular kind of listener will enjoy.

14. Visceralization: sonification beyond data representation

Despite the particular properties that make sonification an opportune means of informational representation, its affordances go beyond replacing or complementing visualization. In this sense, sounds have the potential to influence users' emotions by affecting the pleasantness of an interaction (Serafin et al.). Accordingly, sonification is *data visceralization* where the goal is not only to represent information, but also engage in an emotional and physical interaction with the information: "Visual things are for the eyes, but visceralizations are data that the whole body can experience– emotionally, as well as physically" (D'Ignazio and Klein 84).

Thus, sonification understood as a form of data visceralization is not only about what is heard, but about what is felt —where sounds run through time retaining the liveliness of specific events (Palmer and Jones). In some instances, the musical and analytical uses of sonification converge. For example, Kosara writes of the sonic memorial to the victims of the Pulse nightclub shooting in Orlando in June 2016, in which the display sounds "a note for every one of the 49 victims, for every year of their lives," that while the project is undoubtedly a powerful experience, "[i]t's more of a performance of the data ... not just a representation. Its power is that it touches and moves, not that it gives a clear idea of how many people were killed that night, how old they were, etc."

Singling visceralizes linguistic data, which is inherently a humanist geography, comprising varying degrees of uncertainty, aesthetics and narrativity. Accordingly, and in comparison to print, sonification does not engage in noetic closure, where the meaning is isolated on a written surface and presented as self-contained and complete —rather, it returns utterances to "the world of thought and expression" (Ong 121) where a text is never finalized nor closed: Singling gives back the *incompleteness* to the textual data that has been lost in print. Furthermore, and as discussed by Ballatore, Gordon and Boone in their experimental work on sonifying uncertainty in various fields (information design, geography and data science), sound dimensions such as loudness, order, and clarity can be used to effectively communicate uncertainty in various scenarios.[4]

15. Conclusion: The map is (not) the territory

As discussed above, Singling can be thought of as a metaphorical mapping of language that does not attempt comprehensive mimetic reproduction in the style of Borges's cartographers' in the famous story *On Exactitude in Science*, who map their kingdom inch for inch and mile for mile, so that the map itself is as large as the territory. We are on the fringes of a computational lexicographic project in which cartographers of natural language have taken on the Borgesian challenge of mapping human discourse with such exquisite and fixed precision, that it may be navigated and analyzed predictively and computationally by artificial intelligence. While the goal of Singling is to turn text into data for analysis, our analytic method is not via the other-than-human intelligence of computation, rather it is via the delicate mechanism of the human ear. Deleuze and Guattari write, "The map is open and

[4] For more information on sonification for uncertainty, see also Lohda, Wilson and Sheehan.

connectable in all its dimensions; it is detachable, reversible, susceptible to constant modification. It can be torn, reversed, adapted to any kind of mounting, reworked by an individual, group, or social formation. It can be drawn on a wall, conceived as a work of art, constructed as a political action or as a mediation" (10). If we think of language as a mapping of meaning, and sonification as a sonic mapping of language, then the map is *also* the territory. More finely granular than visualization, sonification nonetheless abstracts the listener from the precise level of semantics. What is lost from specific signifier-signified lexical relations is supplanted with each token's intertextual signification. Our intention is not only to sonify the general meaning that each word or symbol contains, it is also to amplify lexicogrammatical features and listen along the contours of the map: the hills, valleys, street networks, thoroughfares, and deadends, in order to divine new patterns and relationalities within that territory. Importantly, a sonic mapping of language moves beyond the visual, and brings time, duration, and an ephemeral viscerality into the phenomenological experience of the data as terrain.

Yet facility with literacoustics remains remarkably undeveloped among researchers and scholars. The perpetual bias toward observed rather than audited data remains a blindspot of academic and scientific scholarship, even in the era of total datafication and the naturalized synaesthesia made possible by synaesthetic media (Waterworth 1997): if seeing is believing, then it may be that hearing is key to critically questioning and understanding through embodied sympathetic vibrations with data. The value of such a text-to-sound software is manifold. We think that there is value for any researcher who is blind or vision impaired. There is value for the researcher who wishes a deeper experiential resonation with their data, who feel physically distanced and dehumanized by vision's analytic propensity to disrupt continuities, to drain affect and aesthetic excess from the observed, to freeze life and compartmentalize the remains. There is value for those who wish to retain sequential nuances within the mood or tone or voice of a text over time. Visualized singularities (including polysingularities) in the form of a graphical display of textual data show the overall text as an entirety but cannot capture the transitioning text— the text in time where rhetorical and emotive energies reside. There is value in attuning our ears to notice irregularities as opposed to the norms of a text, to hear what makes a text exceptional as opposed, or in addition to, the generic, genre-based features it possesses. Part of the point of text-to-MIDI sonification algorithms is to make linguistic data "unfamiliar again. By translating it, transcoding it, remediating it, we begin to see elements of the data that our familiarity with visual modes of expression [like typeface and the alphabet] have blinded us to" (Graham).

We are still early in the development of the Singling project, which we hope will continue to evolve and proliferate both within the DLC and beyond as further iterations of the code and database are committed on https://github.com/eyung/Singling. To date, we have used Singling for live musical performance, to aid live narrative composition, to accompany virtual poetry reading while simultaneously sonifying the performed poems, as well as for a studio recording by The Hired Minds called "Singling's First Adult Conversation in Five Parts." In exploring the use of this software initially from an arts-based perspective we have not only grown more familiar with the potential transformations performable on any text, but with those sets of parameters we are individually most attuned to. This is not unlike learning to read, in which pleasure drives the acquisition of a skill. We take this moment in the narrative emergence of the interface to map the interwoven and inter-agential pathways we have traveled to get to this instantiation. Now we turn our attention to the specific analysis of textual data having already identified ways to meaningfully and systematically listen to the algorithm's sounding. We share our results across collaborators in tiny properties files and aggregate these with our own settings to enhance collective processes of analysis and mapping of our textual data. Much like a Singling composition, this mapping of our process is neither definitive, fixed, nor singular; rather, it is a kind of sensory posthuman collaboration, a coming together of humans, technologies, algorithms, timelines, and modalities within a community of scholars and artists.

Works Cited

Baccianella, Stefano, et al. "SENTIWORDNET 3.0: An Enhanced Lexical Resource for Sentiment Analysis and Opinion Mining." Proceedings of the 7th International Conference on Language Resources and Evaluation, 2010, pp. 2200–2204.

Ballatore, Andrea, et al. "Sonifying Data Uncertainty with Sound Dimensions." *Cartography and Geographic Information Science*, vol. 46, no. 5, 2019, pp. 385–400, doi:10.1080/15230406.2018.1495103.

Barrass, Stephen, and Paul Vickers. "Sonification Design and Aesthetics." *The Sonification Handbook*, edited by Thomas Hermann et al., 2011, pp. 145–72.

Borges, Jorge. L. "On Exactitude in Science." *A Universal History of Infamy*, translated by Norman Thomas de Giovanni, Penguin Books, 1975, p. 131.

Bourdieu, Pierre. *Distinction: A Social Critique of the Judgment of Taste*. London: Routledge and Kegan Paul, 1984.

D'Ignazio, Catherine, and Lauren Klein. *Data Feminism*. MIT Press, 2020.

Debashi, Mohamed, and Paul Vickers. "Sonification of Network Traffic Flow for Monitoring and Situational Awareness." *PLoS ONE*, 13(4), 2018, doi:10.1371/journal.pone.0195948.

Deleuze, Gilles, and Félix Guattari. *A Thousand Plateaus: Capitalism and Schizophrenia*. Bloomsbury Publishing, 1987.

Drucker, Johanna. "Visualization and Interpretation: Humanities Approaches to Graphical Display." *Digital Humanities Quarterly*, vol. 5, no. 1, 2011, pp. 1-21.

Edmonds, Philip, and Eneko Agirre. "Word Sense Disambiguation." *Scholarpedia*, https://www.scholarpedia.org/article/Word_sense_disambiguation#A_sense_inventory_cannot_be_task-independent, accessed 29 May 2020.

"English Prepositions." *Wiktionary*, https://en.wiktionary.org/wiki/Category:English_prepositions, accessed 29 May 2020.

Fellbaum, Christiane. "WordNet(S)." *The Encyclopedia of Applied Linguistics*, 2012, pp. 1–8, doi:10.1002/9781405198431.wbeal1285.

Ferguson, Jamie, and Stephen A. Brewster. "Evaluation of Psychoacoustic Sound Parameters for Sonification." ICMI 2017 - *Proceedings of the 19th ACM International Conference on Multimodal Interaction*, 2017, pp. 1–8, doi: 10.1145/3136755.3136783.

Graham, Shawn. "The Sound of Data (A Gentle Introduction to Sonification for Historians)." *The Programming Historian*, https://programminghistorian.org/en/lessons/sonification, accessed 29 May 2020.

Halliday, Michael Alexander Kirkwood. *On Language and Linguistics*. Vol. 3, A&C Black, 2003.

Hashemi, Mahmoud. "Hatnote and Wikipedia Projects." *Sedimental: Accretionary thoughts by Mahmoud Hashemi*, https://sedimental.org/hatnote_projects.html, accessed 28 May 2020.

Hermann, Thomas. "Taxonomy and Definitions for Sonification and Auditory Display." *14th International Conference on Auditory Display*, 2008, pp. 1–8.

Illich, Ivan. "A Plea for Research on Lay Literacy." *Literacy and Orality*, edited by Nancy Torrance and David R. Olson. Cambridge University Press, 1991, pp. 28-46.

James, Kedrick. *Toward a Theory of Choice through Visuals*. https://www.researchgate.net/publication/341753649_Toward_a_Theory_of_Choice_through_Visuals, doi: 10.13140/RG.2.2.30740.53121.

_____. "The Piano Plays a Poem." *Poetic Inquiry: Enchantment of Place*, Vernon Press, 2017, pp. 49–58.

James, Kedrick, et al. "The Patch: An Artful Syn(Aes)Thetic Mapping of Linguistic Data through Collaborative Digital/Analogue Literacy Processes." *McGill Journal of Education*, 2021, forthcoming.

Juliano, Linzi, and Ramesh Srinivasan. "Tagging It: Considering How Ontologies Limit the Reading of Identity." *International Journal of Cultural Studies*, vol. 15, no. 6, 2012, pp. 615–627, doi:10.1177/1367877912451684.

Kohll, Oliver. "TextSound1." *GitHub*, 31 Mar. 2012, https://github.com/okohll/TextSound, accessed 26 May 2020.

Kosara, Robert. "Sonification: The Power, The Problems." *EagerEyes*, 2 March 2017, https://eagereyes.org/techniques/sonification-the-power-the-problems, accessed 30 May 2020.

Kramer, Gregory, et al. *Sonification Report: Status of the Field and Research Agenda*. March 2010, https://digitalcommons.unl.edu/psychfacpub/444, accessed 30 May 2020.

Kumbhar, Rajendra. "Modern Knowledge Organisation, Systems and Interoperability." *Library Classification Trends in the 21st Century*, edited by Rajandra Kumbhar, Chandros Publishing, 2012, pp. 95-1113. doi:10.1016/B978-1-84334-660-9.50008-4.

“Letter Frequency.” *Wikipedia*, Wikimedia Foundation, https://en.wikipedia.org/wiki/Letter_frequency, accessed 26 May 2020.

“Lexnames.” *Princeton University*, https://wordnet.princeton.edu/documentation/lexnames5wn, accessed 26 May 2020.

Lodha, Suresh K., et al. “LISTEN: Sounding Uncertainty Visualization.” *Proceedings of the IEEE Visualization Conference*, 1996, pp. 189–95, doi:10.1109/visual.1996.568105.

Mayzner, Mark S., and Margaret Elizabeth Tresselt. “Tables of Single-letter and Digram Frequency Counts for Various Word-length and Letter-position Combinations.” Psychonomic Monograph Supplements, 1965.

Minghim, R., and A. R. Forrest. “Illustrated Analysis of Sonification for Scientific Visualisation.” *Proceedings of the IEEE Visualization Conference*, 1995, pp. 110–17, doi:10.1109/visual.1995.480802.

Moretti, Franco. “Conjectures on World Literature.” *New Left Review*, no. 1, 2000, pp. 54–67.

Neuhoff, John G. “Perception, Cognition and Action in Auditory Displays.” *The Sonification Handbook*, edited by Thomas Hermann et al., 2011, pp. 65–85.

Norvig, Peter. “English Letter Frequency Counts: Mayzner Revisited or ETAOIN SRHLDCU.” *Peter@Norvig.com*, https://norvig.com/mayzner.html, accessed 26 May 2020.

Novak, Petra Kralj, et al. “Sentiment of Emojis.” *Plos One*, vol. 10, no. 12, 2015, doi:10.1371/journal.pone.0144296.

Ong, Walter. *Orality and Literacy*. Methuen & Co, 1982.

“Oxford Languages: The Home of Language Data.” *Oxford Languages*, https://languages.oup.com/, accessed 26 May 2020.

Palmer, Michaela, and Owain Jones. “On Breathing and Geography: Explorations of Data Sonifications of Timespace Processes with Illustrating Examples from a Tidally Dynamic Landscape (Severn Estuary, UK).” *Environment and Planning A*, vol. 46, no. 1, 2014, pp. 222–240, doi:10.1068/a45264.

Princeton University. "About WordNet." *WordNet*. Princeton University, 2010.

Romans, Brian. “Data Sonification.” *Wired*, April 2007, https://www.wired.com/2007/04/data-sonification/, accessed 26 May 2020.

Rosenblatt, Louise M. “The Transactional Theory of Reading and Writing.” *Theoretical Models and Processes of Reading. International Reading Association*, 1994, pp. 1057–1092.

Serafin, Stefania, et al. “Sonic Interaction Design.” *The Sonification Handbook*, Logos Verlag, 2011, pp. 87–110.

Shaw, Debra B. “The Aesthetics of Retrieval: Beautiful Data, Glitch Art and Popular Culture.” *Anthropocenes – Human, Inhuman, Posthuman*, vol. 1, no. 1, 2020, doi.org/10.16997/ahip.15.

Strapparava, Carlo, and Alessandro Valitutti. “Wordnet affect: an affective extension of wordnet.” *Lrec*, vol. 4, 2004, pp. 1083-1086.

The MIDI Association. *The History Of MIDI*. https://www.midi.org/articles-old/the-history-of-midi, accessed 29 May 2020.

Torres, Rui. "Digital Poetry and Collaborative Wreadings of Literary Texts." *New Media and Technological*, 2005, pp. 1–9, http://www.telepoesis.net/papers/dpoetry.pdf.

Venugopal, Vineeth. "Scientists Have Turned the Structure of the Coronavirus into Music." *Science*, 3 Apr. 2020, https://www.sciencemag.org/news/2020/04/scientists-have-turned-structure-coronavirus-music, accessed 26 May 2020.

Visse, Merel, et al. "The Unsayable in Arts-Based Research: On the Praxis of Life Itself." *International Journal of Qualitative Methods*, vol. 18, 2019.

Voegelin, Salomé. *Listening to Noise and Silence: Towards a Philosophy of Sound Art.* Bloomsbury Publishing USA, 2010.

Walker, Bruce N., and Michael A. Nees. "Theory of Sonification." *The Sonification Handbook*, edited by Thomas Hermann et al., Logos Verlag, 2011, pp. 9–40.

Waterworth, John A. "Creativity and Sensation: The Case for Synaesthetic Media." *Leonardo*, vol. 30, no. 4, 1997, pp. 327-330.

Wittgenstein, Ludwig. *Prototractatus: An Early Version of Tractatus Logico-Philosophicus.* Cornell University Press, 1971.

Wolfe, Joe. Note Names, *MIDI Numbers and Frequencies.* https://newt.phys.unsw.edu.au/jw/notes.html, accessed 27 May 2020.

World Heritage Encyclopedia. "Letter Frequencies." *Project Gutenberg*, https://www.self.gutenberg.org/articles/eng/Letter_frequencies, accessed 26 May 2020.

Worrall, David. *Sonification Design: From Data to Intelligible Soundfields.* Springer International Publishing, 2019, doi:10.1007/978-3-030-01497-1.

Chapter 8

The GTR Language Workbench: A History of Creative Computational Collaboration

Sharla Sava

IBM

Abstract: "The GTR Workbench: A History of Creative Computational Collaboration" focuses on the GTR Language Workbench, a web-based platform invented by David Ayre and Andrew Klobucar. Leveraging advances in Natural Language Processing (NLP), the Workbench is an accessible tool for anyone interested in working creatively with language, offering a unique visual interface for collaborative creative writing and text analysis. Surveying the history of the Workbench's invention and development since the nineties, Sava traces the cultural and intellectual influences shaping the founders' approach to technology. Connecting this initiative with relevant touchpoints in avant-garde poetry and machine learning, the author points to GTR as a valuable contributor to the dynamic field of computational poetics.

Keywords: Natural Language Processing (NLP), Object-oriented Programming (OOP), artificial intelligence, machine learning, Tipster, lexical resource, Kootenay School of Writing, Vancouver

***

The vodka is good, but the meat is rotten. [1]

The GTR Language Workbench is a web-based platform, providing a unique visual interface for collaborative creative writing and text analysis. In this

[1] Kumar attributes this to a team of Machine Translation researchers at Georgetown University in the post-WW2 era. (Kumar, 4).

essay, I explore how the GTR Workbench leverages computer processing to augment and improve human expression. Given it's absurd, creative glitchiness, it persistently defies utility. GTR exemplifies the critical contribution avant-garde poetry and experimentation make to the world of computing, including the growing viability of using Natural Language Processing (NLP) creatively, for purposes other than advertising.

This essay takes time to consider the founders within their socio-historical context. This is in part a corrective given that, while GTR has been around for decades and contributed to the computational poetics community, the endeavor has not been systematically archived, documented or historicized. The lack of access to local history makes it difficult to understand what is at stake in both the vision and the output of GTR. The other reason for taking time to explore the history of the project is to avoid a narrative of technological determinism; to emphasize from the outset that human imagination, rather than simply technology, drives this endeavor and is moving us forward into the realm of creative computation.

GTR, an acronym that stands for Global Telelanguage Resources, is the invention of David Ayre and Andrew Klobucar.[2] The founders first met in Vancouver in the early nineties, crossing paths at poetry readings hosted by the Kootenay School of Writing (KSW). Their meeting at the KSW was not a singular occasion so much as a foundational context, as the KSW represents a history of poetic and intellectual ideas that are central to understanding GTR. Klobucar had first learned about the KSW during grad school, where he studied modernist poetry and literary theory, first from Drew Milne (his advisor at the University of Edinburgh), and later Peter Quartermain (his advisor at the University of British Columbia (UBC)). Klobucar became a regular at KSW events through the nineties, while completing his doctorate at UBC on Charles Olson's ecological approach to poetics. Ayre was a computer programmer and had come across the KSW while taking creative writing classes. As he became familiar with KSW poetry, Ayre immediately recognized parallels between programming and the experimental approaches to language typical of KSW poets. When they met, Ayre was studying artificial intelligence as part of his undergrad degree in Computer Science at Simon Fraser University (SFU). Both founders eventually became members of the KSW collective and acknowledge the central importance that the KSW represents in their intellectual and aesthetic work.

---

[2] Personal interview. Many of the historical details about GTR were provided during this interview.

Further insights about the poetic and social approach to writing offered by the KSW can be found in the historical record, including by Klobucar himself. After being involved in the collective for several years, Klobucar would partner with another collective member, Michael Barnholden, to co-edit an anthology, *Writing Class - The Kootenay School of Writing Anthology* (1999). In their introduction, the editors emphasize the central importance of collaboration: "Inspired by labour unions and their capacity to effect significant social change, KSW often worked as well as wrote collectively. The prominence of collaborative work and thinking demonstrates the school's high regard for communal identities" (Barnholden, 5). Neither formalized nor programmatic, collaboration stemmed from a common approach to poetry, rooted in language: "What held KSW writers together has less to do with specific social ideals or shared notions of the communitarian good, than a critical sense of language itself as a prime constituent of community in general. In other words, community operates here as an effect of language, rather than the other way around, as more traditional writing and art scenes might assert"(5). As we will see, this approach, which relies on language as the basis of meaningful collaborative activity, carries through to GTR as well.

1. First Iteration: Global Telelanguage Resources Performs

The first iteration of GTR was ephemeral and performative. In the mid-nineties Klobucar and Ayre adopted a pseudo-corporate identity, imagining GTR as a construct that allowed for conceptual play with technology and language. Ayre had initially been inspired by a summer internship at IBM, where he was struck by the language and style used by corporate culture. At their first live performance at an artist-run center in Vancouver circa 1998, Ayre and Klobucar adopted pseudonyms, Daniel de Souza (Ayre) and Fraser Thomas (Klobucar), dressed in corporate suits, and presented the theory and language of GTR. It was pure satire, and they stayed in character throughout. This kind of performance echoed decades of west coast avant-garde experimentation with alternative identities, from Leopard Lady to N.E. Thing Co.. What differentiates GTR from this tradition of West Coast performance art, however, is the kind of experimentation that surfaced at the level of the machine, and computation.

During that time, Ayre was at Simon Fraser University studying Object-Oriented Programming (OOP) and Artificial Intelligence. He learned LISP and Java programming languages and would eventually adopt Java as his main language. He began to experiment with using computational language models to develop computer programs for creative language play. Rather than simply a clever conceptual performance filled with random language play, what GTR performed was the output generated by their engagement with language software. Beyond satire, computational exploration informed and guided the presentation as a whole, and would turn out to be more enduring and generative.

The enthusiasm and centrality of technology during this moment is, of course, consistent with the dot com era, part of the sweeping cultural transformation brought about by the first internet boom (1995-2001). After graduating in 1999, Klobucar spent a few years working in tech, as an editor at CNET in San Francisco, and Ayre took a Senior Developer role at HYWY Software, a startup in Toronto. They stayed in touch and both ended up returning to Vancouver around 2001. GTR gradually evolved into a theory of instantiation, a framework and project aimed at using software to generate language games with a sense of play typical of Dada and Surrealism. Ayre borrowed the concept of instantiation from OOP, which allows developers to use class definitions to create objects to use in a given program. GTR would keep the grammatical structure of a source text but enable the user to trade out parts of speech (e.g. nouns, adverbs, etc.) from one or more "palette" texts. The approach used a statistical model called term frequency-inverse document frequency (TF-IDF) which measures word importance within a given document/text. While there is no theory of GTR instantiation per se, their Proof of Concept was workshopped in the Vancouver community and shared online for a time. Called "The Instantiator," it was intended as a *service* that poets could use to generate texts infinitely.

2. Second Iteration: GTR Workbench (32-bit desktop)

The second iteration of GTR, then, crystallizes around 2004, when Ayre and Klobucar were invited for a residency at the Banff Centre (Banff Centre, 16). While the transition between phases is characterized by the move from satire to software, it is materially confirmed by the invention of what would become known as the GTR Workbench (32-bit desktop version). The Banff New Media Institute, located spectacularly high in the Rocky Mountains, provides, as explained by their 2004 Catalogue, "the facilities, technical resources and expertise to see projects through to fruition" (Banff Centre, 2). Furthermore, as a co-producer, it "provides a naturally inspiring location in a multidisciplinary arts environment which stimulates creative collaborations and connects artists with the media" (Banff Centre, 2). With a summer away from work and daily routines, Ayre and Klobucar found the time and resources to bring GTR software to life (see Fig. 1). They acknowledge that the preciousness of Banff is also connected to the creative and intellectual inspiration provided by peers from around the globe, and they established what would become life-long relationships with media artists David Jhave Johnston, Maria Lantin and others.

Figure 8.1. David and Andrew's workspace at The Banff Centre (July 2004) Photo: David Ayre (used with permission).

The GTR Workbench, as a desktop application, was an expansion of the Instantiator. It abstracted the specific logic of the Instantiator into the more generic concept of a language "processor," of which the Instantiator would be just one of many. The Workbench would house a variety of these language processors, eventually allowing users to mix and match them to create even more complex multi-processors. While they were at Banff, Ayre and Klobucar used this early version of the Workbench to evolve what they called the Dictionary Project. They used open-source natural language processing databases to build a processor that generated neologisms - not just new words, but also definitions for them. Here is how Klobucar and Ayre described it at the time:

> The Global Telelanguage Resources Workbench functions currently as a smart "word processor/work bench," in other words, a "language toolbox" able to submit any text to a number of creative, transformative algorithms in the form of specially designed filters. At the moment, these filters allow for the spontaneous generation of new words from a wide range of selected databases, complete with their own customized etymologies and definition and translation extensions. The tool will

> eventually make possible an unlimited number of literary and aesthetic modifications to texts, much as current graphic design software like Photoshop and audio editors like Sound Forge permit artists to create and modify image and sound files (Klobucar 2006).

The chain of textual transformations within the Workbench was modeled on the Unix operating system's "pipeline" concept. Imagine adding text as the input to one end of the pipeline. The text travels through the pipeline and encounters a mini-computer program, or language processor. As it enters the first processor the text might be analyzed and transformed by a sorting processor that is concerned with identifying line length. It exits the first processor and continues along inside the GTR pipeline system. The next processor the text encounters might be a synonym replacement that relies on recognizing noun patterns. Processed again, the original text will be transformed a second time. Eventually, the text, released from all the processors, continues through the pipe until it exits as output, or new text. The pipeline metaphor is useful because it is modular; it describes the way a text can be transformed a little or a lot, depending on how many processors it encounters in the pipeline chain before exiting the GTR system.

Today, in the online version of the GTR Workbench, the Instantiator still appears as one of many language processing tools. To give you a sense of what it does, here is an example. I used a speech by former President Donald Trump as the source text, and then stirred in, if I recall the ratio correctly, 50% nouns and 70% adjectives from a book chapter from the astrophysicist Neil deGrasse Tyson, and here is a bit of what came back:

> For military remains, Kim cohabited a kind of
> human soup in which
> the easy ban continually scattered the people of higher frequency.
> Back then, if your law had
> been seen across China's court, you couldn't have traveled.
> Any people entering your shorter wavelength would, just
> nanoseconds earlier, have bounced
> off the immigrants right in front of your Mexico. You would have seen
> only glowing wall in
> all directions. They were fantastic,
> simple and lethal in color.

The text, undoubtedly intentionally, inspires questions about authorship and disrupts accepted ideas about the nature of creative practice. Whose text is this? Was it written by a human or a machine? Is it accurate for me to call myself the author? Is this original? Can it be called creative? While the author in a traditional sense may be impossible to locate, I have no doubt that my own attempts at poetry benefit from having the Instantiator as a personal language coach, even co-author.

Another one of the language processing tools that Ayre and Klobucar developed for the GTR Workbench is the Travesty Generator. Based on Travesty, a computer program originally co-created by Hugh Kenner and Joseph O'Rourke and published in BYTE Magazine in 1984, the generator (both in 1984, and in the GTR Workbench) was designed to scan texts and generate a "nonsense" imitation. As Klobucar points out in a 2012 essay, the results were inevitably humorous: "Regardless of how familiar (or even readable) the resulting words might be, consistent with both the name of the ensuing text and the program itself, each generation proudly announces itself as a 'travesty.'" Klobucar continues, pointing toward the literary and aesthetic critique prompted by this type of computer poetry: "Kenner and O'Rourke are also emphasizing its inherent absurdity, reminding readers that whatever combinations may result from their programming, the new texts continue to misrepresent and destabilize all source texts in the form of mockery—works in search of new patterns and possibly missed or overlooked structures (Klobucar 2012). Kenner and O'Rourke, like Klobucar and Ayre, represent a dynamic convergence between humanities and computer science. Based at Johns Hopkins University at the time they published in BYTE, Kenner was known for his scholarship on Ezra Pound, and O'Rourke as a professor of computer science in the field of computational geometry. There is room to explore why academic scholarship on modernist poetics, earlier with Kenner on Pound and later on with Klobucar on Olson, appears to lend itself to the production of computer poetry.

In terms of application development, the GTR Workbench leveraged advances in artificial intelligence, more specifically in the domain of Natural Language Processing (NLP). Ayre was able to leverage the work done by the DARPA-funded Tipster project. Active in the nineties, Tipster was a research and development group whose goal was to advance text processing capabilities for the intelligence community, specifically in the areas of document detection and information extraction (Merchant). Ayre came across the documentation, released openly, and used the Tipster architecture framework as the core of the GTR Workbench.

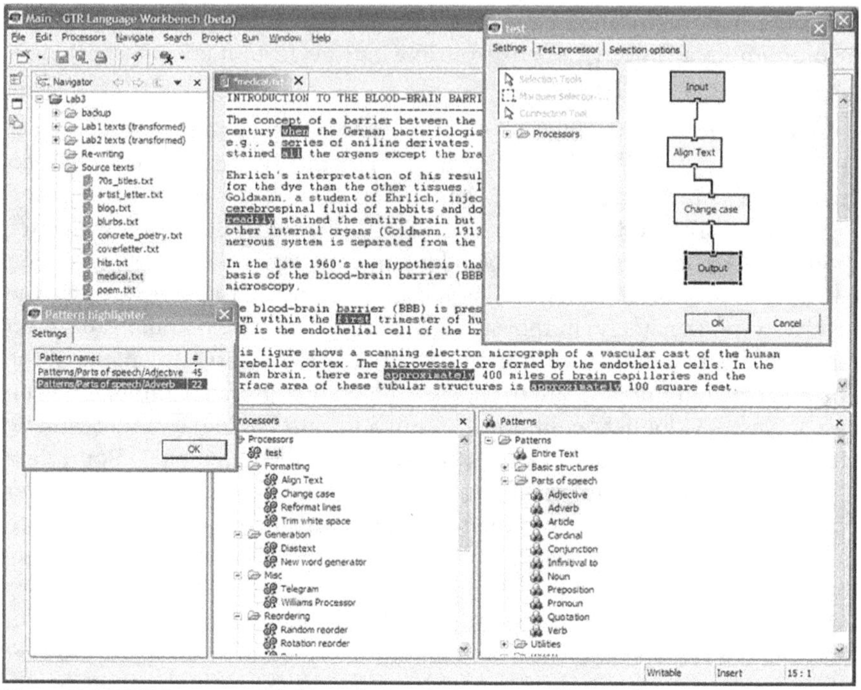

Figure 8.2. GTR Language Workbench (beta) circa 2012. Photo: Chris Funkhouser (used with permission).

It's crucial to understand that GTR makes an original contribution when it comes to coding NLP software (see Fig. 2). Ayre envisioned and programmed architectural elements on top of the Tipster framework to enable users to explore the creative and poetic aspects of language. While some of the processors, like Travesty Generator, are intentional integrations of existing code, others are built from scratch. GTR contains algorithms created specifically to digitize avant-garde literary practices, including Dadaist and Surrealist word games. And both the Instantiator and the Williams Processor, a mini-program that transforms text input to writing in the style of poet William Carlos Williams, are unique to GTR.

In the years following Banff, community interest in GTR expanded. Ayre and Klobucar continued to develop the Workbench, with periodic experiments and shout-outs appearing online.[3] Although the GTR Workbench, as a desktop application, was functional, Ayre and Klobucar would face increasing

[3] References include Cone, Funkhouser, Johnston, Eddeaddad (Antonio Roque).

challenges maintaining compatibility with various operating systems. During that era, following 2008, further disruptions to the project occurred when Klobucar relocated from Vancouver to the New York area to take an academic position at NJIT and Ayre left his day job as a programmer, built a tiny house on a remote west coast island, and went off-grid.

The founders stayed in touch, and a new vision for GTR began taking shape. By about 2013 it had become clear that the digital environment allowed for more sophisticated web-based applications hosted in the cloud, solving the OS compatibility and software maintenance issues. Another factor at the time was Klobucar's proximity to the dynamic startup community in New York City, which provided constant opportunities to exchange ideas and witness emerging public-private partnerships and technology trends. The vibrant ecosystem was supportive of academic innovation and startup ventures and it was easy to find like-minded thinkers to exchange ideas about online education, software advances, and emerging business models. Venture capital, present across industry sectors, allowed the ecosystem to thrive, creating endless opportunities for meetups, demos, hackathons and accelerators. While energized by the startup culture at the time, GTR steered clear of adopting the model. Faced with the aggressive requirements posed by venture-capital agreements predicated on the need for constant user expansion and monetization, Klobucar and Ayre declined to turn GTR into a for-profit business, retaining control of project development in service of the poetry community.

3. Third Iteration: GTR 1.0

The third iteration of GTR, a cloud-based version of the application that would be accessible via any web browser, came into focus around 2016. Ayre and Klobucar released their Beta version of what is now called GTR 1.0, at the Electronic Literature Organization's 2016 conference, held at the University of Victoria (Next Horizons). In addition to the processors built for the original desktop version of the Language Workbench, the roadmap for GTR 1.0 envisions a number of expanded elements. While Workbench users had engaged with the Workbench individually, by 2016, Ayre and Klobucar were able to imagine how to leverage new software capabilities to allow users to connect and share their creative work through a shared site library. They continued to share their progress with the computer poetry community and run GTR workshops in a variety of locations, presenting "The Community and the Algorithm: A Digital Interactive Poetics" at the ELO17 conference in Portugal (Torres, 2017), the CCCC conference in Missouri (2018), and when Klobucar was a visiting fellow in the Department of Linguistic, Literary and Aesthetic Studies at the University of Bergen, in Norway (2018). Klobucar presented the keynote for the LLED Graduate Student Conference in Vancouver (2019), and both founders ran a "Digital Poetics Workshop" hosted by The Capilano Review in Vancouver (2019) (see Figs 3,4,5).

Figure 8.3. David Ayre (left) and Andrew Klobucar (right) presenting at CCCC (Kansas City, Missouri, March 2018) Photo: Sharla Sava (used with permission).

For the past several years, Ayre has continued to develop the backend of GTR 1.0, updating the original Tipster framework, improving the reliability of the processors for users, and increasing the efficiency of connecting to the various NLP linguistic databases used by the GTR system. In 2018, Ayre began the development of an API based server and the creation of a unified database to store the main vocabulary datasets, merging WordNet 3.0, CMUDict and The SPECIALIST syntactic lexicon. For those unfamiliar with NLP tools, some background is helpful. Wordnet is "a large lexical database of English" that is freely available. Its "structure makes it a useful tool for computational linguistics and natural language processing" (Wordnet). CMUDict, "The Carnegie Mellon University Pronouncing Dictionary is an open-source machine-readable pronunciation dictionary for North American English that contains over 134,000 words and their pronunciations. CMUdict is being actively maintained and expanded" (CMU). The third database, The SPECIALIST syntactic lexicon, is "a large syntactic lexicon of biomedical and general English, designed to provide the information needed for the SPECIALIST Natural Language Processing System. Coverage includes both commonly occurring English words and biomedical vocabulary. The lexicon entry for each lexical item records syntactic, morphological, and orthographic information. The lexicon has been released as one of the UMLS Knowledge Sources since 1994" (Lexical Systems Group). The cloud configuration relies on containerization using both Docker and Kubernetes.

Early in 2020, under the direction of Professor Osama Eljabiri, the Ying Wu College of Computing at NJIT provided a new opportunity for supporting the development of GTR 1.0. The Capstone Program, as it's known, partners computer science seniors with business sponsors for a semester-long work-study placement (see Fig. 6). The Capstone, providing students with an opportunity to solve real-world problems and sponsoring the time to work with skilled and motivated students, enabled Ayre and Klobucar to speed up progress on the backlog items on their roadmap. While the project had to be curtailed as a result of shut-downs triggered by COVID-19, it was meaningful while it lasted, expanding their collaborative practice into a new context.

Directing GTR toward the needs of digital humanities educators (both K-12 and Higher Ed), Ayre and Klobucar are also working on curated collections, based on Workbench language processors, which they call Language Packs. For example, the "William Carlos Williams Language Pack" contains a biography, a stylized lexicon of key terms used by William, links to additional resources, and GTR processors that allow users to create in the style of the poet. Language Packs can be integrated into an existing curriculum as modules that include learning activities, lesson plans and teacher guides. Additional planned features include a mobile-friendly UI and the creation of a drag-and-drop interface (reminiscent of visual programming tools like MIT's Scratch) that will allow users to easily select and combine processors. Another component on the roadmap, more on the technical side, allows users interested in coding their own processors to use the Workbench's JavaScript-based API.

4. Conclusion

GTR's embrace of NLP as the basis for creative practice suggests that computational poetics is poised to make a valuable contribution to the field of artificial intelligence. NLP, originally known as computational linguistics, is commonly traced back to the 1950s, with the field of machine translation and early efforts to use computers to translate between human languages. There is the popular example, likely spurious[4] but nevertheless widely cited, of an amusing translation between English to Russian that starts with the aphorism "The spirit is willing but the flesh is weak," and ends up as "The vodka is good, but the meat is rotten" (Kumar, 4). Amusing as an anecdote, it offers a way of understanding the degree to which computers were thought to be estranged from natural language or informal conversational styles.

I imagine it's fair to say that not long ago, certainly within living memory, the primary function of computers was understood to be that of mathematical calculation. Moreover, language, connected to cognition and intelligence, was more comfortably situated in the realm of human communication. In his wonderful 1963 book, Jeremy Bernstein effectively captures the moment. Describing his relationship to computers as a theoretical physicist, he explains the challenge in terms of a conversational struggle, pointing out that humans can't speak directly to machines: "For one cannot simply present a problem to a machine in English, and expect it to come up with the answer; the computer has a language of its own - machine language - and whatever one wants to tell it must be translated into that language" (5). Decades later, this is no longer the case. Creative initiatives such as GTR present writers with the ability to mine linguistic databases that a single human, even across the span of a lifetime, would simply not have time to synthesize. As computers rely on statistical modelling and machine learning techniques to learn from massive databases of human words, identified by grammatical and syntactical tags, Bernstein's notion of translation between human and machine is slipping from focus. GTR is not a machine that removes human agency, a generator that usurps the human voice and turns poetry into a cold, dry computational process. It's an invitation to join an expanded ecosystem where humans and computers are engaged in a creative process that never ends.

Works Cited

2018 CCCC Convention Program. https://cccc.ncte.org/cccc/review/2018program Accessed 18 May 2020.

Ayre, David and Andrew Klobucar. Personal Interview. 17 May 2020.

Banff Centre, The. Banff New Media Institute Co-production Catalogue. 2004. www.debraswack.com/pics/bnmi_co-pro_2004.pdf Accessed 9 May 2020.

Barnholden, Michael, and Andrew Klobucar. *Writing Class: the Kootenay School of Writing Anthology*. New Star Books, 1999.

[4] John Hutchins argues that this is a persistent myth. For more, see "'The whisky was invisible', or Persistent myths of MT" *MT News International* 11 (June 1995), pp.17-18.

Bernstein, Jeremy. *The Analytical Engine: Computers – Past, Present, and Future.* Random House Inc., 1963.

Capilano Review, The. "Digital Poetics Workshop with Andrew Klobucar and David Ayre" May 2019. https://thecapilanoreview.com/payment-confirmation/digital-poetics-workshop-with-andrew-klobucar-and-david-ayre/ Accessed 5 May 2020.

CMU. The CMU Pronouncing Dictionary. http://www.speech.cs.cmu.edu/cgi-bin/cmudict Accessed 11 May 2020.

Cone, Tom and Jeremy Todd. "Collaboration and Constructivism" Emily Carr Institute of Art and Design Public Lecture Series. Fall 2006 https://www.monroco.com/downloads/ECIAD%20Lecture%20Series.pdf Accessed 5 May 2020.

Eddeaddad (Antonio Roque) "other tools: GTR Language Workbench, Rhyming Robot (seuss), Infinite Monkeys" *Gnoetry Daily,* 2 August 2011. https://gnoetrydaily.wordpress.com/2011/08/02/other-tools-gtr-language-workbench-rhyming-robot-seuss-infinite-monkeys/ Accessed 6 May 2020.

Funkhouser, Chris. "On Using Tools Made by My Comrades" *Netartery,* 17 June 2010. http://netartery.vispo.com/?p=146 Accessed 7 May 2020.

Hutchins, John. "'The whisky was invisible', or Persistent myths of MT" *MT News International* 11 (June 1995), pp.17-18.

Johnston, David Jhave. "Interview with Andrew Klobucar" *elmcip* 7 February 2012 (video). https://elmcip.net/critical-writing/interview-andrew-klobucar Accessed 8 May 2020.

Klobucar, Andrew. "Articulate Filters, Fluent Bodies: Reading Alan Sondheim's Writing Under" *Hyperrhiz 15.* Fall 2012. http://hyperrhiz.io/hyperrhiz15/essays/klobucar-sondheim-writing-under.html Accessed 2 May 2020.

Klobucar, Andrew. "Artifice & Intelligence - New Writing, New Technologies." *The Capilano Review,* Fall 2006. https://thecapilanoreview.com/issues/fall-2006-artifice-intelligence/ Accessed 1 May 2020.

Kumar, Ela. *Natural Language Processing.* I.K. International Publishing House, 2011.

LLED Graduate Student Conference Program. University of British Columbia, May 2019. https://blogs.ubc.ca/lledgradconference2019/files/2019/05/Program-Abstracts-19.05.01-Vers.-2.pdf. Accessed 19 May 2020.

"Lexical Systems Group - The SPECIALIST Lexicon." (2020) *National Library of Medicine,* https://lexsrv3.nlm.nih.gov/Specialist/Summary/lexicon.html. Accessed 12 May 2020.

Merchant, Roberta H. "Tipster program overview." Proceedings of a workshop on held at Fredericksburg, Virginia: September 19-23, 1993. https://dl.acm.org/doi/pdf/10.3115/1119149.1119151 Accessed 3 May 2020.

Next Horizons 2016: Electronic Literature Organization Conference & Media Arts Festival. https://dtc-wsuv.org/projects/elo2016site/. Accessed 15 May 2020.

Princeton University. "WordNet: A Lexical Database for English." WordNet. Princeton University. 2020. https://wordnet.princeton.edu/#:~:text=WordNet%C2%AE%20is%20a%20large%20lexical%20database%20of%20English.&text=Synsets%20are%20interlinked%20by%20means,and%20publicly%20available%20for%20download. Accessed 10 May 2020.

Chapter 9

I Am Afraid: Voice as Sonic Sculpture

Maria Lantin and Simon L. Overstall

Emily Carr University

Abstract: This Chapter discusses a multi-user networked VR application, I Am Afraid, which offers a simple, visually calm space to externalize thoughts into interactive objects in 3D space. Each object encapsulates the sound of the voice which conjured it and can resonify this voice in response to actions such as scrubbing, shaking, looping, and repositioning. Objects can be spelled out words, or abstract shapes typically mapped to non-word vocalizations. The actions of participants in the space create an ever-evolving audiovisual sculpture of words and other vocalizations made strange through real-time effects, loops, and overlaps. I Am Afraid has been used for fun, performance, and therapeutic purposes. It provides a poetic space for engaging with text as sculpture and instrument.

Keywords: Virtual Reality, Performance, Sound, Sound Sculpture, Sonic Objects

***

1. Introduction

It has been said that sounds are as close to us as our thoughts (Berkeley, sect. 51). The inception of the virtual reality application *I Am Afraid* was an exploration of remixing thoughts as externalized entities. What if we could see and hear thoughts and play with them as virtual instruments? How would this transform their affective power?

We developed a multi-user networked VR application, *I Am Afraid*, which uses voice as an interface to create sonic objects in a virtual environment. Words are spoken and added to the environment as three-dimensional textual objects. Other vocalizations are rendered as abstract shapes. The sculptural elements embed the sound of the voice that initiated their creation and can be played as instruments via user-controlled interactions such as scrubbing, shaking, or looping. Multiple users can simultaneously be in the environment, mixing their voices in an evolving, dynamic, sound sculpture.

The first instantiation of this idea took place within the digital puppetry theatre show *Manipulation* (Pepper's Ghost) where the performers shouted words solicited from the audience and made them appear as three-dimensional textual objects on a screen behind the performers. The words were then used as virtual instruments and played back by the performers: forwards/backwards, slow/fast, singly/overlapped (see Figure 9.1).

Figure 9.1. *Cloud Opera* performance in *Manipulation* digital puppetry piece presented at the Vancouver Fringe Festival, 2016. Words appear on the screen after being spoken and can be interacted with a motion-tracked stick. In this figure, the character on the right of the image has the stick which can play back the word "Tadaa" on the screen (Photo by Alan Goldman, © Maria Lantin).

The theatre performance demonstrated that this type of word manipulation could be playful and engaging. The next instantiation was in the context of a TEDx talk on the topic of fear for which we developed the first version of the *I Am Afraid* (IAA) Virtual Reality (VR) app (see Figure 2). The speaker entered the stage, donned a Google Daydream headset, and initiated her talk by performing the emotion of fear using a pre-generated sound sculpture composed from fear-induced words (e.g. "what's going on", "I like it", "I don't like it", "why did he do that") (See Figure 9.3).

Figure 9.2. Performance with I Am Afraid VR App. Maria Lantin speaking at TEDxECUAD and performing with the VR application *I Am Afraid.* March 2017 (Lantin) (Photo by Jerez Challenger on tedxecuad flickr, CC BY-NC-ND 2.0).

2. Implementation Details

The original version of *I Am Afraid*, VR edition, was built for the Google Daydream platform using the GVR SDK and the Unity gaming engine. Because the Google Daydream platform has since been discontinued, a newer version for the Oculus Quest is in development. The app is a multi-user networked application which has been tested with up to three users at once, though the number of users could be increased with a strong stable wi-fi network. Participants are represented as masks, reminiscent of neurons (See Figure 9.6). All interactions use a controller with one button that can also act as a trackpad. The interaction metaphor is that of a laser pointer which triggers different actions when intersected with the sound objects. The audio can either be played back through headphones or shared with an audience through speakers. However, the latter creates a more difficult recording environment and influences the range of sonic sculptures that can be created.

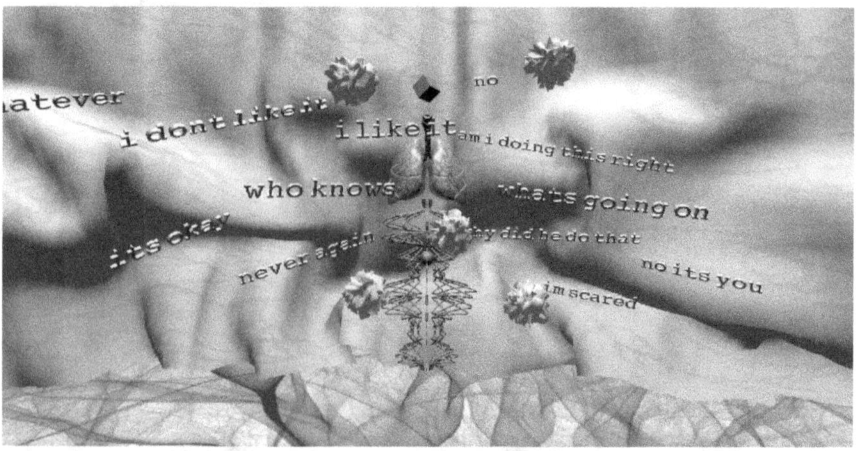

Figure 9.3. Screenshot of participants in VR application *I Am Afraid* Participants in the networked multi-user VR application are represented as masks akin to neurons.

A cloud-based Speech to Text service is used to transform a spoken phrase into a textual object in real-time. The speech recording is stored on a server and fetched by all the clients as soon as the object is created in the shared environment. The newly created object contains the sound of its creation and can be played back using the controller. The playback of words is done through granular synthesis whereby the sounds are broken apart into small chunks and recombined in real-time (Roads). Moving the laser pointer on a word object allows the performers to scrub through the word's sound at any speed, forward or backward, and at an arbitrary position. The volume of a sound object can be changed by moving the object higher or lower in the space, allowing for a more complex mix of sounds.

Non-word sounds, or sounds that prefer to resist textual representation, can also be added to the environment. These are represented as objects similar to paper balls and can be played back in whole or in part by moving the laser pointer onto the paper ball, which loops the sound until the pointer exits the object. The non-word sounds are instrumental in creating a textured soundscape. Some of the more interesting sonic sculptures created in IAA have made extensive use of a "library" of paper balls.

More complex playback is possible by creating loops of sounds which then continuously play and can be layered with other loops or interactive sounds. A loop is created by "right-clicking" and drawing a path on and around a sound object (currently, only one object can be involved in a single loop).[1] The path then loops continuously or until a participant turns off the loop by right-clicking again. A small red sphere retraces the path while in playback mode, in a comet-like fashion, which lends another playful dynamic element to the sonic sculpture being developed (See Figure 9.4).

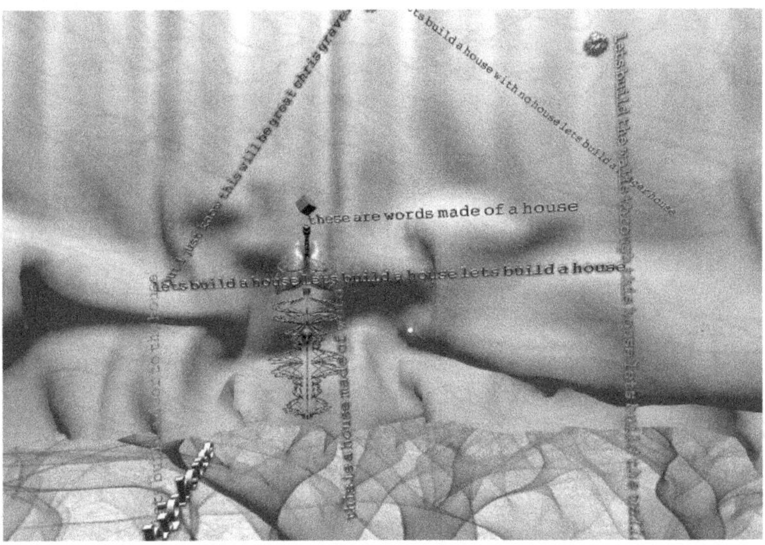

Figure 9.4. Screenshot of *I Am Afraid*, VR Edition. The words and objects highlighted in yellow are looping on a custom path. The red spheres are following the path.

[1] "Right-Clicking" here refers to pressing the single button/trackpad on the controller with the finger on the right side of the button/trackpad. Other active regions are the bottom, and center.

Deleting a word explodes it into its component characters which then linger on the floor. Deleting a paper ball shrinks it to nothing. Words can also be moved and rotated to create a dynamic sculptural and sonic composition.

3. Sound and Poetics

The recent resurgence of Virtual Reality has not, thus far, produced many interactive apps that are focused on sound. Even in apps whose focus is sound composition, the environment is designed for visual stimulation, akin to music visualizers. Scott Niedjadlik's project *SoundscapeAR* (Niedjadlik) uses objects and paths in space to compose music in augmented reality using the Hololens platform. While it does not use voice, it does create a low-key, yet engaging interface for composition and playback. In the VR app, *InTone*, Scott Hayden uses voice intonation to influence the movement of graphical object clusters (Hayden). The voice is not specifically tied to the graphical elements but does promote a synesthetic association. In-world music composition is also an emerging genre that allows the mixing of different instruments and sounds that are mapped to 3D objects in space. *LyraVR* is a notable example in this genre that integrates spatialized and playful elements for an embodied composition experience (LyraVR).

Virtual Reality as it is currently offered, even in apps that are ostensibly about sound, privileges visual perception. We literally place a screen two inches from our face. This overemphasis on the visual is perhaps because of a long-standing bias equating presence with seeing and seeing with knowledge. However, sound is in many ways closer to virtuality than image because, with more tenuous material referents, it can have a more diffuse affect. In effect, soundscapes are the original virtual reality. The visual design of IAA is deliberately low-fi, with grey tones, little movement, and simple geometries that loosely reference interiority and the body (See Figure 3). It is meant to be a counterpoint to many virtual reality experiences which overwhelm with hyperreal, saturated, constantly changing images. The intention is to foreground the experience of creating a sculptural soundscape from voiced thoughts. The interaction proceeds with simple gestures that generate a layered spatialized sound environment through loops and direct playback. The result is an iterative feedback loop of listening and doing, a collaborative phenomenal dance with sound. The experience has been described as meditative, engrossing, playful, and potentially endless, which reflects the way immersive sound interaction can transform our usual way of perceiving the world.

Figure 9.5. Screenshot of the Virtual Application, *I Am Afraid.* A series of phrases and sounds on the topic of fear were recorded and placed in the virtual space. Each is embedded with the sound of the recording and can be reperformed.

The spatial construction of the textual objects and paper balls offer a visual poetic space that augments the sonic performance. The 3D dynamic sculpture offers an anchor to the meandering subjects of the performance. As with written poetry, it is possible to link structure to subject to reinforce or break meaning. Figure 9.6 shows a composition that evolved into a house-like structure. We have witnessed many poetic strategies employed, including nonsensical words deliberately elicited from the Speech-To-Text engine (the Dada strategy), short words that can be composited in different ways (the fridge magnet strategy), and haikus and odes (the Formalist strategy). Spatially, the strategies varied widely depending on whether movement or visual effect was being emphasized. Often, when a new performer discovered that shaking and moving a word produced a new sonic possibility, the structures became more chaotic and re-configured as spatial poetics in a moment to moment delight.

In general, the intermixing of visual and sonic representation of words creates possibilities that are experienced potentially quite differently for author and audience, as both are re-interpreting and remixing sensory attention in real-time.

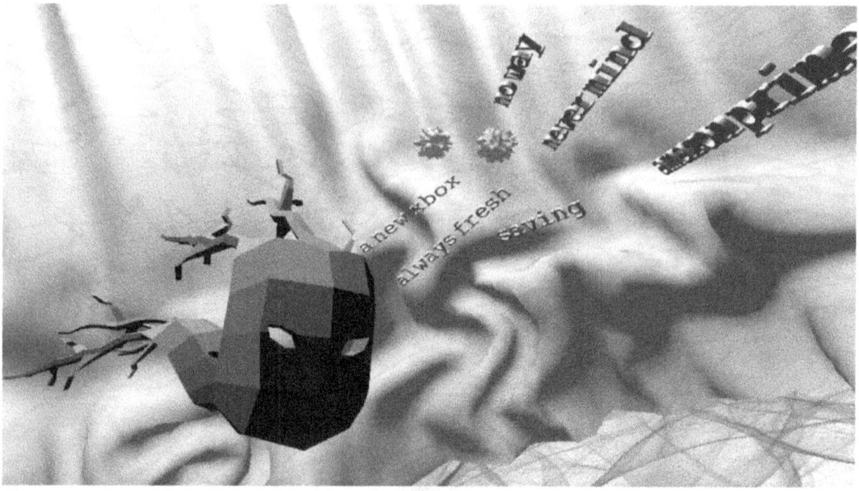

Figure 9.6. A performance about building houses made of words made of houses made of sounds.

4. I Am Afraid in Performance

In addition to the TEDx talk, IAA was used as a performative tool for a conference presentation (Lantin, "AR in Action") where input from the audience was sought and integrated into the app via clowning techniques. The end of the talk featured a remix of the voices of the audience. In this particular case, IAA was used as a way to externalize the collective thoughts of an audience and was meant as a provocation or catalyst for further thought during the conference. We have found that the use of an AI-driven Speech-To-Text service is an unlikely but worthy clowning partner as it often returns unexpected results that can make for spontaneous moments of improv with machine-generated puns.

IAA was also used as an additional performative element in *Tangles and Bloom* a dome show integrating Pablo Neruda's poem "Ode to the Gillyflower" (62) with interactive animation at the Société des Arts Technologiques in Montréal in May 2018 (Lantin, see Figure 9.7).

Other plans for performances include the use of the application, or concepts from the application, for live narrative interpretations, and dance improvisation.

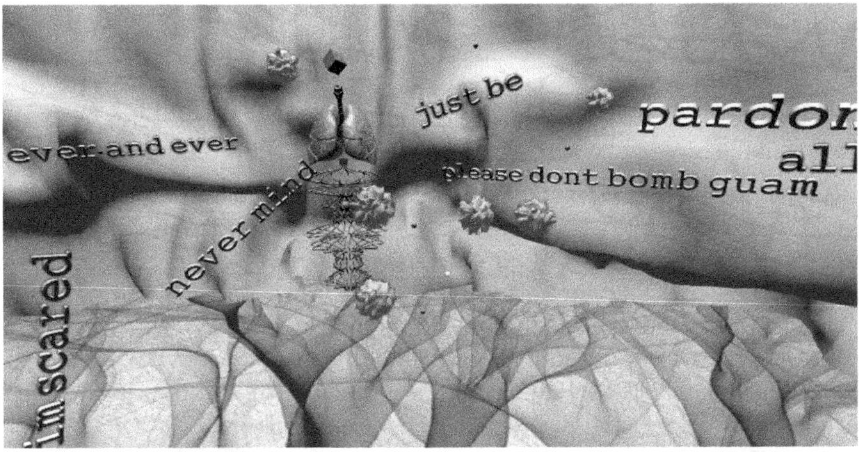

Figure 9.7. Documentation of Tangles and Bloom performance in May 2018, Montréal, QC. The two performers are in headsets with a VR version of the dome that includes interface elements to modify the dome show for the audience. In the later stages of the animation, the performers add words based on the poem to the dome and play them back (Photo by Tom Kuo, used with permission).

One of the first skills taught in clowning is connecting with the audience in a direct and vulnerable way. Using a Virtual Reality application for a performance that includes clowning is awkward because the headset can be seen as a barrier between the performer and the audience. We have found a few ways to work with this barrier, such as taking the headset on and off as a performative gesture (like a scuba mask) and integrating the VR environment as an internal world that is shared with the audience using naïve techniques (i.e. the Auguste clown). Some of these limitations will be mitigated when lightweight Augmented Reality (AR) glasses are more widely distributed. The ability to see each other while performing will promote the sense of shared space with the audience and will lend a sense of materiality to the words in space. Indeed, our first experimentation with IAA in the Oculus Quest, which allows passthrough video and positional tracking, has shown that the nature of the interaction between performers changes towards more collaborative spatial designs. In the Google Daydream version of IAA, the performers' positions are static, and all interactions are done via a kind of laser pointer. With the addition of positional tracking, it is possible to walk around the words and to create compositions that are position-dependent.

5. Users as performers

We have demonstrated IAA in many different venues and have noticed that a sense of play develops between participants in the app (up to three could be in the same, networked, environment), and between participants and their audience. As words and vocal sounds, from different voices, build up in the environment, there is a sense of call and response or dialogue with present and past participants. There can also be a playfulness in deleting sounds which, in the case of words, explode into individual letters onto the floor.

In one demonstration instance (Lantin, "IX"), we had an unexpected visit by some Baroque musicians who happened to be visiting from Germany for a concert performance. Without knowing their background, we put them into the IAA environment and were delighted at the layered, harmonious, yet chaotic soundscape they produced together. They employed a strategy of creating a library of harmonized sounds and words, which they then remixed, dynamically creating loops and turning them on and off in an extended jam session. This highlights the role of prior experience and virtuosity in any new instrument. Our own experience has shown that, with practice, the space of sound possibilities expands far beyond what IAA's simple interface might suggest.

6. Fear and Virtuality

An investigation of fear was the predominant motivation for this VR app, which was being developed while the 2016 U.S. Election campaign was underway. During this time, there was a marked increase in anxiety in the general population and words were very much at the centre of both the source and expression of that anxiety. Experientially, fear is felt in the body and one of its key secondary characteristics is discursive, often repetitive, thought. In "Smile at Fear," Trungpa argues for a kind of giving up of privacy, a radical look at what is actually happening (21) in order to cut through the negative consequences of fear states. It occurred to us that transducing a bombardment of thoughts from an internal monologue space to an external, interactive, networked virtual space could shift the experience from claustrophobic to playful and shared.

Thoughts exist within the virtual field of awareness, constantly appearing and disappearing without a definable or volitional source. Virtuality as defined by Deleuze foregrounds this potentiality and constant change (97). The stories we tell ourselves come from a field of potential that enacts the next moment and change the stories. In this sense, it is impossible to depict the virtual, to externalize it, because, as soon as we do, it slips away. According to Massumi, "this does not, however, preclude figuring it, in the sense of constructing images of it. To the contrary, it requires a multiplication of images. The virtual that cannot be felt also cannot but be felt, in its effects. When expressions of its effects are multiplied, the virtual fleetingly appears. Its fleeting is in the cracks between and the surfaces around the images (133).

In creating an intermediate space between the virtual and the actual, we invoke a space to reflect on the background of fear, the field that manifests it. It is, as Levy describes, the "objectivation of interiority and the subjectivation of exteriority," (18) – a way to re-virtualize our thoughts playfully. Levy further articulates the effect of virtualizing text and the reading process itself as a way of actualizing text in infinitely more ways than we could with the printed page. It has deterritorialized text. He points out that "it has created a text that lacks sharp outlines or a definable interior. Now there is only text, as we might say water or sand. The text has been set in motion, swept up in the flow, vectorized, metamorphosed. In this sense, it is much closer to the movement of thought or its representation" (62). The particular kind of hypertextualization created by IAA is a conduit linking thought to image and sound in a kind cycle between interiorization and exteriorization that Levy likens to a Möbius strip.

Experientially, the Möbius effect is felt most directly and immediately in IAA upon hearing one's own voice made strange. We become alienated from our own thoughts in a way that questions their meaning, somewhat like the feeling of repeating a word over and over again until it is no longer a word but just a series of mouth movements and sounds. In further remixing the words or partially embedded sounds, there is a shift from the rumination of the fear state to an exploratory (and later intentional) creative space.

7. Future Work

The roadmap for IAA is based on the development of a comprehensive set of physically-based sonification engine of virtual words and objects. Some additions planned for the future include:

- throwing words/objects;
- knocking words/objects together;
- stretching, quashing, and squeezing;
- sending out streams of particles that activate the words/objects they hit; and
- words shaped around a curve or onto another object.

The design of the current environment is specific to a VR headset without position tracking. All performers appear next to each other, facing a centrally located sculpture with breathing lungs. The position of the performer is fixed, and only head movement is possible and visible to other performers. The sculpture is intended to anchor the space into a particular orientation and to act as an interaction object (e.g. the blue cube is for recording words, and the red wire body is used to record non-textual sounds). In our first tests of the IAA environment in a headset that includes positional tracking, the central sculpture appeared out of place and cumbersome. Our first tests also indicated that much of the interactivity will need to be redesigned for a more embodied performance. The Oculus Quest SDK now includes hand location and pose tracking which could enable a controller-free interaction. Being able to reach out and touch or move through sound objects with others will likely bring further virtualization of thoughts, enhancing their potential outcomes toward the performative and playful. Our tests indicated that performers were intuitively drawn to the possibility of bumping into words as a means of playing them, either with their heads or hands. It is quite possible that even with these additional embodied gestures, there would still be a need for a laser-based interaction for a more zoomed out performance that could incorporate more objects/words in quick succession. In one version of the app, we may use pass-through video to create the feel of an augmented reality application where the physical room is visible, and the virtual words/objects are added as overlays. In the context of a public demonstration, this is often desirable as many are reluctant to enter into a vulnerable state where they can be observed without their knowledge. Ultimately, different versions will need to exist to support co-located performances when a shared physical space is possible, and remotely networked performances where only a virtual space can be shared. It is possible that in the case of Augmented Reality some of the intimacy of the fully immersive environment will be lost. In our experience, augmented reality does direct a certain kind of relationship with objects, which is less virtual, and more actual. However, the primacy of sound blended with movement may introduce another kind of intimacy, that of co-creating a musical dance in space.

We are also looking at adding adaptive algorithms to the app that could intervene in the sonic sculpture, such as conjuring objects based on interactions (e.g. knocking two words together) or spontaneously adding words from a chosen corpus of words/documents. In general, building a mixed reality app in support of a poetic generative sound space can go in many different directions. At each iteration, we gain a great deal from witnessing a wide range of performers using and misusing the app.

8. Conclusion

I Am Afraid is an evolving project which seeks to externalize or re-virtualize thoughts into sonic sculptures. It started as a fascination with words and their embedded hidden sounds, and subsequently intersected with a long-standing investigation of fear and its manifestations in the body. We have created a collaborative, immersive space where spoken words become instruments and sculptural building blocks, and eventually invitations to dance and perform with others. The emphasis on sound creates an intimate and meditative space, providing a counterpoint to the mostly visual spaces that are currently available in Virtual Reality. The interface is embodied and mostly invisible to promote a performative and virtuosic experience for both performer and audience.

We see IAA as a new kind of hypertext that brings in new ways of writing, reading, and performing text. The newly enabled interactions with the text blend interior and exterior conceptual spaces in a dynamic loop that generates meaning from spatial, visual, and audio elements. Making music of words and sounds is an ancient community-based practice. IAA aims to take this practice into virtuality and towards infinite actualizations.

Acknowledgements

This research was supported by the Social Sciences and Humanities Research Council of Canada, Mitacs, Canada Foundation for Innovation, and Emily Carr University of Art + Design.

Works Cited

Berkeley, George. *An Essay Towards a New Theory of Vision.* 2003. *Project Gutenberg,* http://www.gutenberg.org/ebooks/4722.

Deleuze, Gilles. *Bergsonism.* Zone Books, 1988.

Hayden, Scott. *Sing To The Stars in VR.* https://youtu.be/xeatsFrfipA. Accessed 28 Jan. 2018.

Lantin, Maria. *AR Sound-Mixing | Maria Lantin | AR in ACTION.* 2017, https://youtu.be/IbbK3ox8kBg.

_____. *I Am Afraid.* http://ix.sat.qc.ca/2017/programmation-liste?language=en. IX Immersion Experience Symposium: Embodied Spaces, Montreal, QC, Canada.

_____. *I Am Afraid | Maria Lantin | TEDxECUAD.* 2017, https://youtu.be/2xB_3F1bIEI.

_____. *Tangles and Blooms.* CONNEXIONS (CHI 2018), Société Des Arts Technologiques (SAT), Montreal, QC, Canada.

Lévy, Pierre. *Becoming Virtual: Reality in the Digital Age.* Plenum Trade, 1998.

LyraVR Inc. "LyraVR." *LyraVR. lyravr.com,* https://lyravr.com/. Accessed 28 July 2020.

Massumi, Brian. *Parables for the Virtual: Movement, Affect, Sensation.* Duke University Press, 2002.

Neruda, Pablo, et al. "Ode to the Gillyflower." *Odes to Common Things,* 1st ed, Bulfinch, 1994, pp. 62–67.

Niedjadlik, Scott. *SoundScapeAR Spatially Aware Audio Mixing & Visualization | Scott Niejadlik | AR in ACTION.* 2017, https://youtu.be/QAaNVjdeRkg.

Pepper's Ghost New Media & Performing Arts Collective. *Manipulation.* Vancouver Fringe Festival, Vancouver, BC, Canada.

Roads, Curtis. *Microsound.* MIT Press, 2001.

Trungpa, Chögyam, et al. *Smile at Fear: Awakening the True Heart of Bravery.* 2010. *Open WorldCat,* http://www.myilibrary.com?id=960806.

Chapter 10

Afterwords: As We May Enact

Anna Nacher

Jagiellonian University, Poland

Abstract: The current pandemic crisis provides an important opportunity to look critically and creatively at the many technological options now in play to support collaborative, online communication. This paper holds fast to the instant we are all experiencing, newly fearful and subsequently reticent to engage in normal face-to-face interactions, whether professional or personal, due the coronavirus. Nacher's reflections look at her own recent explorations in digital collaboration, while frankly admitting that her relationship to the online screen might best be described as a flash of "hope in the dark." Clearly, communication is changing with or without the virus, and she is "still figuring out what it means to think in synch" with her colleagues and friends using the platforms available, along with their governing algorithms.

Keywords: COVID-19, Collaborative Writing, Writing Platforms, Media Ecologies

***

On a dark and gloomy late February evening, I was sitting in front of my desktop computer after having absolved three consecutive 90-minute online class meetings, the online 40-minute faculty meeting, followed by the online administrative 2-hour meeting for the doctoral program I am chairing, barely able to keep my eyes open, with my mind apparently wrapped in a fog as dense as the winter air pollution outside my window. Yet, still curious and intrigued as we may think (thank you, Vannevar) in sync while writing collaboratively - even if hit hard by a Zoom fatigue - I was waiting to connect to a monthly ELO Virtual Salon gathering, initiated in 2019 by Deena Larsen and Annie Abrahams.

1. Beyond Zoom fatigue

The idea of such a gathering started circulating around the time of the last year's 2019 ELO Conference in Dublin, followed by a broader discussion on how to proceed with the initial idea that was carried out in the late summer and autumn of that year. The first meeting was organized in January 2020, so the very beginning of the series coincided with the start of the pandemic lockdowns in Europe (February – March 2020) and the U.S. (April – May 2020). Each subsequent Salon has since followed a different format. I remember particularly well the first meetings in February when Kiril Azernyi inspired us to collaboratively and "electronically" read and interpret Nika Skandiaka's experimental poem. Also memorable was the Salon in March, featuring Annie Abrahams and her fascinating "reariting" of Lily Robert-Foley's "Extra-terrestrial Rhetoric," offered as a collaborative performance of community improvised writing with Framapad. Then the May's Salon led by Reham Hosny was dedicated to Amira Hanafi's "A Dictionary of the Revolution." I owe the details to the generous and consistent archiving effort by Johannah Rodgers, available at this point as a Google Document. The first three Salons prompted me to wonder how collaborative thinking becomes collaborative writing and vice versa? Even if – or, especially when – "writing" is not necessarily "inscription"? Granted, there are traces on the platforms (much like smoke on the water, considering the lack of stability and corresponding obsolescence of some of the corporate and commercial cloud computing accounts), and almost every move we make gets datafied (often without our will, although with our consent). Yet, there is still some meaningful "between."

In most cases, such an improvised, collaborative thinking/writing in synch happens between platforms and services, with Zoom as a central node. However, it does not mean relegating the other digital components to the periphery, far from it. Collaborative, improvised writing with open source and open access Framapad proved to be an ecstatic experience also at the 2020 ELO conference where it constituted the important element of "(un)distance" panel proposed by Annie Abrahams and the team (including Deena Larsen, Roderick Coover, Kiril Azernyi, Renee Carmichael, Johannah Rodgers, and Anna Nacher). Framapad is not the part of corporate, predominantly English-speaking internet of platforms. It was built by Framasoft, the French non-profit organization that calls to "degooglize your Internet" and is committed to building "the emancipatory digital world" (Framasoft), mostly through offering free and transparent online services (it is no coincidence that using either version of Framapad, setting up the profile is not required). Each of them bears Frama- in its name, including Framabook, the publishing house; Framalibre, the free culture directory or Framablog, the blogging platform. I remember my first contact with this francophone collaborative notebook in 2013, while participating at the Digital Humanities Summer School at the University of Bern, where several dozen participants used it to collectively make notes during lectures. In the meantime, Framapad has outgrown its initial server capacity and currently consists of several "pads." Comparing it with Google Docs reveals a few minor but significant differences, including the necessity to set up an account with the latter, opening the users to inevitable data mining and profiling by the company that has succeeded in establishing a late-capitalist equivalent of ancient emporia. On the other hand, Framapad does not allow extra-textual commenting. When it comes to the collaborative enterprise of ELO's Salons, the already mentioned archiving effort should not be forgotten, as it is crucial for further exploration, research, and work in the future, especially in the case of improvised online reading / writing sessions. Facebook and (sometimes) Twitter also constitutes an element of the Salon's broader media ecology (via announcements and comments accompanying content posted in the Electronic Literature Organization's group), but remains rather peripheral in terms of collaborative writing.

2. Restructuring with the post-

To what extent can collaborative writing serve as an answer to many nowadays crises? What happens when an algorithmic entity – however ambiguous and entangled in the logics of corporate digital capitalism and intensive, profit-and metrics-oriented data tracing – becomes the part of a treasured relationship? How is it that our future has vanished all of a sudden? Is this evacuated future worth mourning to start with? I am forced to ask these questions, while browsing my 2019 and 2020 calendars. It would be tempting to say that a significant change occurred around mid-March 2020, when the country where I am based went into a full lockdown, as one of the preventive measures in the EU-wide attempt to fight the COVID-19 pandemic. My notes, however, clearly demonstrate that the transformation in question at that point had already been in the making. The process apexed with the decision to withdraw my individual paper from 2020 ELO Conference, which was motivated by an increasing difficulty to write as a sole author in the working environment already saturated with collaborative endeavors.

By June and July of 2020, most of my intellectual and scholarly work was already happening in-between, through, and because of numerous algorithmically organized conversations, on various platforms, both corporate and open-access: Zoom, Webex, MS Teams, Google Drive and Framapad. A prominence of digitally organized working spaces became noticeable a few months before the pandemic with the significant upsurge in international scholarly and social activity of Electronic Literature Organization member base. Parting our ways after the conference has become more difficult every

year. The change apparently was brought about by the increased willingness to continue the conversations and exchanges that had started, not only in lecture halls and classrooms, but also in cafeterias, passageways, parks, and sometimes even the restrooms of various universities located on different continents. Eventually, it led to questions asked during the process of moving 2020 ELO Conference entirely online: how to re-create with available online tools and platforms all the spaces, both official and unofficial, that thrive on the tactile, physical, and embodied sense of proximity; how to replace/reconfigure/hint at closeness originating from ripples initiated by synchronization of individual breathing rhythms in laughing together, instigated and propagated by bodily micro-movements, punctuated by lines of sight in face-to-face exchanges, and often sealed with a biochemical dance of pheromones. If there ever was an era to celebrate the Deleuzian-Guattarian body without organs, this post-COVID age where we mourn the loss of shared physical presence seems particularly well-suited. Certainly, the notion of a post-COVID age might seem as controversial as its many other counterparts in what has been dubbed "postal deluge" (Blas, 87). Far from denoting any kind of linear and straightforward periodization, though, post- "drifts between eras, politics, peoples, and practices, while it still designates periods and passages of time" (Blas, 87). In its historical reflexivity, with a twist enabling transversal realigning and rearranging practices, the concept seems then an apt vehicle to explore memories that allow us to rediscover and reconfigure all the generative potential of interconnected being – invited and uninvited, intentional and unintentional alike. With all justified skepticism related to almost habitual employment of many instances of postal deluge considered, the notions of post-COVID and postpandemic – if decoupled from a straightforward and reductionist logic of linear periodization – still offer a refreshing vantage point to capture the current moment, even if painful, and force us to come to terms with immense suffering on a global scale. Or, precisely because of that, if we follow another observation on the overabundance of posts: "'Post-' announces that challenging instances of passage and transformation can be articulated through what they proceed." (Blas, 87). And challenging it is - with all the Zoom fatigue, eyes sore from overexposure to blue light emitted from many a screen on a daily basis, many hours a day, including during weekends when we connect to family and friends, in often desperate attempts to maintain at least a specter of warmhearted presence of other human beings and mutual care, despite a grim reality of platforms capitalizing on every aspect of our everyday behaviors possible, including longing for intimate relationships with significant others who happen to be based on the other side of a country, continent or the ocean. What is even more challenging, as eloquently noticed in the chapter by Taeyoon Choi and Iretiolu Akinrinade in the same volume, is how we are going to define community in the world so much dependent on technology to foster and maintain relationships, if the technology in question affords only the particular set of abilities or favors only the particular codes and protocols and not the other, leading to a chain of exclusions not to be easily overcome, forgotten or glided over. The restructuring moment emphasized by Choi and Akinrinade can be seen all the more compelling, when coupled with the educational crisis amplified by the pandemic – as has been widely noticed, COVID-19 amplified many existing crises rather than creating entirely new ones – which can be boiled down to the often inadequate reaction to the necessary shift to teaching online. Anastasia Salter has aptly remarked that for the most part, the focus is still on deployment of new technologies to deliver educational content with fairly traditional methods and the moment of opportunity to radically restructure what online teaching could mean seems to be lost. The latter would, however, require more than the old-fashioned forms repackaged for digital consumption, which is where pedagogy based on critical making advocated by Salter leads.

3. COVID-19, or when the future has evaporated

Thus, it is from such a vantage point that I am shedding a bit of light on some aspects of collaborative, digitally organized, and networked writing spaces that became a staple of professional and private life over the last months. These are often framed by and interwoven with the whole set of crises, including a future that has quite literally vanished. Attending to the collapsed hopes, plans, dreams and agendas from the perspective of this future unmade and equally imploded might be an interesting exercise in crisis management. The future has suddenly vaporized, following a slightly Derridean hauntologic semantic gesture exercised once by Marisa Olson, who famously noticed how offline "vapors" of internet shape her lived and experienced reality, stemming from the fact that the internet by then had become "given" or ubiquitous.

But this rupture occurred much more abruptly, almost from one day to another, decisively signaled by what happened to my 2020 wall planner. For the past few years already I needed the sizeable sheet of laminated whiteboard surface that was usually fulgurated with deadlines, meetings dates and names of the cities within the range of European flight network connecting my hometown with the rest of the continent, marking consecutive trips at the rate of up to 5-6 per month. So, at the very beginning of the pandemic lockdown in mid-March of 2020, I noticed how my future literally evaporated, emptying the consecutive months out of places and events. I erased all of them myself and never bothered to add anything else anymore. Under much-touted "new normal," they are being now automatically added to my various calendars (Google Calendar for my household's shared dates, iCal for the whole Apple circuit, Microsoft Outlook at my Microsoft-dependent university), all synchronized across several platforms. Rushing in a cab to catch the late evening plane to Copenhagen or very early morning flight to London, or taking the tram to avoid morning rush hour and getting on the train to Warsaw has now been replaced with collaborations across time zones, physical states and dimensions, and accompanied with automatic synchronization of time differences, assisted with the World Clock Meeting Planner, whenever the organization of any event involves people from at least three continents (and usually it does).

And that is still just on the surface of COVID19-related breakdowns, which at the end of the day, as I already mentioned, turn out to be accelerated rather than instigated by the spread of the virus, itself, into a global network of transcontinental exchanges and localized versions of public health care failure. The pandemic has exacerbated rather than started the current crisis, where pretty much everything concerning my academic activity and career, how I work, how I communicate with the world, and finally, where all my creativity is being questioned. In a way, it might be seen as business as usual, the sort of questions we - as humanities scholars - ask ourselves quite regularly. After all, for quite some time already, one could notice a growing awareness that such seemingly private questions are often accompanied with the whole set of interwoven social, political and ecological crises, also long in the making, and that academia is neither immune to nor separate from them. To some extent, the history of the modern university can be reframed as sets of answers given to consecutive epistemological shifts, instigated by social and political upheaval often related to significant ruptures brought by changes in media technology, starting with the Kantian idea of university where reason and power would be unified due, in part, according to David M. Berry, "to the rise of print culture, growing literacy and the kinds of destabilizing effects that this brought" (6). In a similar vein, as Berry suggests, digital humanities can be then seen as a necessary attempt to address the "computational turn."

It seems that located right in the prismatic, iridescent center of current turbulences is the very idea and practice of collaboration. Collaborative endeavors happening in networked digital environments, often mediated by increasingly popular platforms, prone to high-level automation and heavy data mining, and resulting in and enabling the traffic being increasingly overtaken by the "custodians of the internet" (Gillespie) and based on "algorithms of oppression" (Noble) have drawn particularly rich and multifaceted critiques. And deservingly so. In many areas of our connected lives, we can see ruins of past dreams about the digital interconnectedness making a difference - with post-Arabic Spring excitement of digital civic engagement epitomized by infamous "Twitter revolutions," too often fading into corporate, for-profit data profiling or, on the other hand, so easy that almost habitual and equally futile "slacktivism" (Tufekci; Schradie). As noticed in relation to "net nationalism" and the policies of censorship so powerfully implemented by the Chinese government, on the occasion of the recent ban of TikTok and We Chat that, regardless of who exerted it, can be perceived as an "overdue" response to such policies, "[t]he idealists who thought the internet would automatically create democracy in China were wrong" (Wu). Not to forget that sometimes the dream of the internet as an inherently liberating technology has brought instead unspeakable, violent responses from respective authoritarian regimes to what was so naively perceived as the scent of freedom enabled by a new communication technology for rapid circulation of news and opinions is no less tragic. We must not forget, in other words, how the consequences of instantaneous and rapid digital circulation are mistakenly taken elsewhere as deliberation – as demonstrated by widespread and increasingly tragic ramifications of conspiracy theories and disinformation narratives du jour. Closer to (scholarly) home: collaborative endeavors seem increasingly trapped within the Digital Humanities taken as a neoliberal "corrective to the "traditional" and outmoded approaches to literary study" (Allington, et al.), which, however, at the same time can sometimes lead to a conservative backlash and return to positivist methodologies excluding rich and vibrant theoretical background gender or postcolonial studies (or, to put it straight and simple, "the neutralization of critique" (Noble, "Toward a Critical Black Digital Humanities") serving the exclusion of research questions asking about power regimes or unwritten cultural assumptions, including sexist and racist, behind algorithmic black boxes. The debate surrounding Digital Humanities as such and as related to emerging subdisciplines is rich, vibrant and nuanced, including a recent significant attempt at examining cross-pollination with electronic literature (Rettberg and Saum-Pascual, Grigar and O'Sullivan). There is no shortage of feminist and racial in-depth critique (Losh and Wernimont; Gold and Klein; Gallon; Risam) and the more nuanced, interest in holding space for the interpretative skill of critical thinking, even if primarily focused on textual analysis, critical Digital Humanities is being proposed (Dobson).

4. Collaborations of the post-pandemic cyborg offspring

Yet, for all unfulfilled promises and seemingly failed dreams of digital activism and for all the disadvantages of neoliberal turn haunting digital humanities, however cliché it may sound, there is still hope – and no, I am not going to write "even if" in the dark. This kind of hope in the dark (Solnit) particularly fits our reality considering the circumstances. Moreover, the (untold) histories with possibilities multiply, proliferate, and flourish. Granted, they may seem to be scattered, passing clouds on an otherwise gloomy firmament of terrifying news, they may lack a unifying theme, and be provisional, ephemeral, prone to get wiped out and vaporized quickly in the indigestible news cycle pumped up on social media steroids. They may seem anecdotal, but nevertheless are extremely pleasant and engaging, just like any version of the famed hero's journey narrative we as humans crave so much. Most often than not, the histories with possibilities come from unexpected and improbable developments, as demonstrated by the surprising results of political engagement of K-pop fandom that intervened in a recent "hashtag war" and effectively silenced the anti-Black Lives Matter enunciations on Twitter or registered in droves for a Trump rally in Tulsa deliberately didn't show up, using their TikTok base to spread the message (some even suggest this was the direct background of Trump's mentioned decision to ban TikTok in the U.S.) (Lorenz et al.). The difference between "now" (the age of the internet of platforms and post-Cambridge Analytica's growing defiance of social media) and "then" (the happy days of naïve trust in democratizing powers of the internet and its early communitarian ethics) is that nobody in their sane mind would even try to come up with any generalizations either about TikTok[1] or about the internet as such.

Are we more willing to celebrate momentary affinities with unexpected allies which may soon fade into obscurity? Or are we so desperate that any sense of community, however short-lived, makes do? Or are we just enjoying the moment of genuine political prank without paying much attention to possible outcomes while holding few, if any, expectations? The thing is, all, some, or none of the above may hold true. Even if contradictory. The very brief history of expectations concerning digital collaborations encourages yet another take and yet another attempt with the hope in the dark, which primarily means realistically touching the ground (of algorithmic black boxes, algorithms of oppression, for-profit digital profiling, dataveillance) while making all kinds of contingent collaborative efforts. If, as Berry reminds us, the "industrial internet," based on a fusion of "algorithmic devices, algorithmic systems and algorithmic decision-making" with physical reality, leads, indeed, to "softening the boundaries between human and machine" (Berry, "Critical Theory and the Digital" 7), then we are already occupying cyborg positions. Saying this does not

---

[1] One might note here that the platform's wild popularity among teens as a social media outlet contributing to progressive causes despite its origins in the autocratic state of the Republic of China resonates with more than one paradoxical storyline and generates an almost infinite number of contradictory interpretations.

mean the celebration of a famous cliché, but rather reminds us of consequences. Among the ramifications of such a hybrid position is a situatedness in all its non-innocent partiality, as "they [cyborgs] are illegitimate offspring of militarism and patriarchal capitalism, not to mention state socialism" who, however, are often "exceedingly unfaithful to their origins" and "completely without innocence." They "do not dream of community on the model of organic family" (Haraway, 151). Which is also to emphasize that in my regular online encounters I am "completely without innocence" and "exceedingly unfaithful" (so well aware of Zoom's privacy policies and safety issues, yet having to come to terms with any doubts, if I am to enact sensible digital collaboration and maintain relationships).

I am still figuring out what it means to think in synch and how we are enacting collaborative writing with my group of like-minded souls, which includes using the services and platforms, to some extent hijacked and repurposed to our ends, contributing uneasily and "without innocence" to our unfaithful community. We write/think in a seemingly precarious and momentarily manner, but stable enough to thrive for decades (thanks to our Great Archivists), resilient enough to secure (un)told histories with possibilities for further explorations. We hope without expectations, possibly wearing "no future" badges (reminiscence of adolescent acts of bravado) and having been grounded in relationships forged while writing together. This idea I owe to Andrew Klobucar and our conversation that kickstarted the very idea of this essay and which does not get referenced in the works cited because lively and inspiring exchange exceeds the notion of bare references and the MLA Style 9th Edition does not include the instance of Zoom conversation (yet).

Works cited

Allington, Daniel et al. "Neoliberal Tools (and Archives): A Political History of Digital Humanities." *Los Angeles Review of Books*, 1. May 2016, https://lareviewofbooks.org/article/neoliberal-tools-archives-political-history-digital-humanities. Accessed 20 July 2020.

Blas, Zach. "Contra-Internet Aesthetics." *You Are Here. Art After the Internet*, edited by Omar Kholeif, Cornerhouse, 2014, pp. 86-96.

Berry, David M. Critical Theory and the Digital. Bloomsbury Academic, 2014.

Berry, David M. "The Computational Turn: Thinking About Digital Humanities." *Culture Machine*, vol. 12, 2011, pp. 1-22.

Dobson, James E. Critical Digital Humanities. The Search for a Methodology. University of Illinois Press, 2019.

Framasoft, https://framasoft.org/en/ Accessed: 18.08.2020.

Gallon, Kim. "Making a Case for Black Digital Humanities." *Debates in the Digital Humanities 2016*, edited by Matthew K. Gold and Lauren F. Klein, University of Minnesota Press, 2016, pp. 42-49.

Gillespie, Tarleton. *Custodians of the Internet. Platforms, Content Moderation, and the Hidden Decisions That Shape Social Media.* Yale University Press, 2018.

Grigar, Dene and James O'Sullivan, editors. *Electronic Literature as Digital Humanities.* Contexts, Forms, and Practices. Bloomsbury, 2021.

Haraway, Donna. "Simians, Cyborgs, and Women." *The Reinvention of Nature.* Routledge, 1991.

Lorenz, Taylor et al. "TikTok Teens and K-pop Stans Say They Sank Trump Rally," *The New York Times,* 21 June 2020. https://www.nytimes.com/2020/06/21/style/tiktok-trump-rally-tulsa.html Accessed 10 Aug. 2020.

Losh, Elizabeth and Jacqueline Wernimont. *Bodies of Information. Intersectional Feminism and Digital Humanities.* University of Minnesota Press, 2018.

Noble, Safiya Umoja. *Algorithms of Oppression. Howe Search Engines Reinforce Racism.* New York University Press, 2018.

Noble, Safiya Umoja. "Toward a Critical Black Digital Humanities." *Debates in Digital Humanities 2019,* edited by Matthew K. Gold, and Lauren F. Klein, University of Minnesota Press, 2019. https://dhdebates.gc.cuny.edu/projects/debates-in-the-digital-humanities-2019. Accessed 10 Aug. 2020.

Olson, Marisa. "On the Internet, No One Knows You're a Doghouse", *e-flux Architecture,* 31. Jul. 2017. https://www.e-flux.com/architecture/post-internet-cities/140712/on-the-internet-no-one-knows-you-re-a-doghouse/. Accessed 7 June 2020.

Rettberg, Scott and Alex Saum-Pascual. "Introduction: Electronic Literature as a Framework for the Digital Humanities", *Electronic Book Review,* 2 Aug. 2020. https://electronicbookreview.com/essay/electronic-literature-frameworks-for-the-creative-digital-humanities/ Accessed 15 Aug. 2020.

Risam, Roopika. *New Digital Worlds. Postcolonial Digital Humanities in Theory, Praxis, and Pedagogy.* Northwestern University Press, 2019.

Schradie, Jen. *The Revolution That Wasn't: How Digital Activism Favors Conservatives.* Harvard University Press, 2019.

Solnit, Rebecca. *Hope in the Dark. Untold Histories, With Possibilities.* Haymarket Books, 2016.

Tufekci, Zeynep. *Twitter and Tear Gas: The Power and Fragility of Networked Protest.* Yale University Press, 2017.

Wu, Tim. "A TikTok Ban Is Overdue." *The New York Times,* 18 Aug. 2020. https://www.nytimes.com/2020/08/18/opinion/tiktok-wechat-ban-trump.html?action=click&module=Opinion&pgtype=Homepage. Accessed: 18 Aug. 2020.

Author Bios

Iretiolu Akinrinade is a social scientist long enamored with the impact of digital media norms across diverse. Having received her bachelor's degree in psychology from Tulane University's School of Science and Engineering amid the 2020 COVID-19 pandemic, she has drifted to the study and critique of practices surrounding digital media technologies through poetic computation and social theory.

Maria Aladren is a stage director, activist, and educator. MCVTS Theatre and exploring a variety of pedagogical approaches to increase innovation in young minds, in collaboration with McCarter Theatre and NJIT. MA (Washington University).

Charles Baldwin is an established digital literary theorist, poet and fiction author with numerous publications. He is currently a director in Strategic Resource Development & Grant Management at Monroe Community College in Rochester, New York.

Taeyoon Choi is an artist and educator based in New York and Seoul. He co-founded School for Poetic Computation in 2013 where he continues to organize classes and teach experimental workshops. He is inspired by the poetics in science, technology, society, and human relations. He works with computer programming, drawing, and writing, in collaboration with fellow artists and community members. He believes in the intersectionality of art, activism, education. He works with activists and scholars on disability rights, environmental justice, and anti-racism. His projects, participatory workshops, performances, and installations were presented at the New Museum, Whitney Museum of American Art, Los Angeles County Museum of Art, Smithsonian Asian Pacific American Center, Van Alen Institute, M+ Museum, and more. He participated in the Shanghai Biennale, Seoul Mediacity Biennale, Istanbul Design Biennale, and Biennale Architettura 2021. He was an artist in residence at Eyebeam Art and Technology Center, Lower Manhattan Cultural Council, Frank-Ratchye Studio for Creative Inquiry, Pioneer Works, NEW INC, Data and Society, and Centre for Heritage, Arts and Textile. He spoke at Creative Time Summit, Decentralized Web Summit, Google SPAN, Adobe 99U, and various academic conferences. He taught at New York University and Parsons School of Design. Through his diverse practices, he seeks a sense of gentleness, magnanimity, justice, solidarity, and intellectual kinship.

Rachel Horst is a PhD student in the Department of Language and Literacy Education at the University of British Columbia. She is interested in 21st century literacies and futures literacies as conceptualized through a Deleuzian ontology of difference. Her work explores the confluence of digital technologies, writing-as-becoming, and narrative futuring for doing theory and cultivating the future(s) imaginary. Informed by decolonial discourse, diffraction, and affirmative difference, she engages with arts-based and experimental research practices that are calibrated towards the ethico-onto-epistemological and aesthetic nature and poetics of qualitative research and data (non)representation.

Daniela Côrtes Maduro holds a master's degree in Anglo-American Studies and a PhD in Materialities of Literature from the University of Coimbra (Portugal). As a researcher, she collaborates with several digital archives and projects focused on the study of narrative, digital media, multimodality, experimental literature and electronic literature. She was a postdoctoral fellow at the University of Bremen (Germany), where she completed a project about electronic literature, and currently works at the Centre for Portuguese Literature (group "Digital Mediation and Materialities of Literature") hosted by the University of Coimbra.

Kedrick James is a Senior Instructor and Deputy Department Head of the Department of Language and Literacy Education. James is a spoken word and multimedia artist, network theorist, teacher and academic researcher. His work applies concepts of procedural poetics and cultural recycling to words, sound and images. He was a classically trained musician who gave up that path to travel the world.

Andrew Klobucar is a literary theorist and teacher specializing in internet research, electronic literature, and semantic technologies. His writings on experimental literary forms and genres continue to analyze the increasingly important role technology plays in contemporary cultural practices in both print and screen formats. He has worked on developing software for writing instruction and written on the use of programmable media in classroom instruction. He is Associate Professor of English at New Jersey Institute of Technology in Newark, New Jersey.

Maria Lantin is Director of the Basically Good Media Lab at Emily Carr University of Art + Design. She holds a PhD in Computing Science (Dalhousie University and Simon Fraser University). She as a director at Banff New Media Institute's Advanced Art and Technology (A.R.T) labs.

Simon Lysander Overstall is a computational media artist, and musician/ composer from Vancouver, Canada. He develops works with generative, interactive, or performative elements. He is particularly interested in computational creativity in music, physics-based sound synthesis and performance in virtual environments, and biologically and ecologically inspired art and music systems. He has produced custom performance systems and interactive art installations that have been shown in Canada, the US, Europe, and China. He has also composed sound designs and music for dance, theatre, and installations. He has an MA in Sound in New Media at Aalto University in Helsinki, a BFA in Music Composition from the School for Contemporary Arts at Simon Fraser University, and an Associate in Music (Jazz) Diploma from Vancouver Island University.

Esteban Morales is a Ph.D. student in the Language and Literacy Education program at the University of British Columbia. He obtained a MA in Educational Technology and Learning Design from Simon Fraser University and a second MA in Transmedia Communication from EAFIT University, Colombia. His doctoral research explores peace education as a lens to understand, critically reflect, and act on social media violence in his home country, Colombia. His research interests include technological mediation, peace education, data and media literacies, and educational technology.

Anna Nacher is an Associate Professor of digital culture at Jagiellonian University, member of the Board of Directors of the Electronic Literature Organization, 2020 Fulbright alumna, an editor of *Electronic Book Review* electronicbookreview.com Latest publication: "#BlackProtest from the web to the streets and back: Feminist digital activism in Poland and narrative potential of the hashtag," *European Journal of Women's Studies,* first published November 2020.

Anastasia Salter is an Associate Professor of Games and Interactive Media at the University of Central Florida, and the Director of Graduate Programs for the College of Arts and Humanities. Dr. Salter is the author of *Toxic Geek Masculinity in Media* (Palgrave Macmillan 2017, coauthored with Bridget

Blodgett), *Jane Jensen: Gabriel Knight, Adventure Games, Hidden Objects* (Bloomsbury 2017), *What is Your Quest? From Adventure Games to Interactive Books* (University of Iowa Press, 2014), and *Flash: Building the Interactive Web* (MIT Press, 2014, coauthored with John Murray).

Sharla Sava is an experienced Ed Tech leader focused on the design and delivery of online learning. With a doctorate in communication and passion for instructional design, adult learning, content strategy, leadership training and learning platform implementations, she has worked at The Cooper Union, Rutgers, and New York Life. She currently works as a learning designer at IBM.

Rui Torres is Professor of Communication Sciences at University Fernando Pessoa, Portugal. He is the director of the book series Cibertextualidades (UFP Press) and one of the editors of the Electronic Literature Series (Bloomsbury Publishing). An author of electronic literature, his poems were published in several anthologies, CD-ROMs and digital archives. He is the coordinator of the Digital Archive of Portuguese Experimental Poetry (https://www.po-ex.net) and a member of the Board of Directors of the ELO - Electronic Literature Organization. His poems and writings are available at https://www.telepoesis.net

Rob Wittig, Assistant Professor, Graphic Design at the University of Minnesota, Wittig has worked for years in the graphic design and publishing world, at one time heading the R&D unit of the Chicago educational publishing firm, Ligature. The focus of Rob's creative research is visual/verbal narrative in electronic media. He is the co-founder of IN.S.OMNIA, a literary electronic bulletin board system that pioneered the creative possibilities of the web. He worked in Paris with Jacques Derrida and Jean-Francois Lyotard, studying technical, artistic, and theoretical aspects of creating visual/verbal literary works with online publishing technologies

Effiam Yung is employed at the Department of Language & Literacy Education in the Faculty of Education, UBC in the role of Web Communications Specialist. He works closely with the Digital Literacy Centre team where their interest in technology and digital media aligns with his. As a programmer, Singling was a fun and interesting project to work on with the DLC members.

Index

D

E

F

G

H

I

J

K

L

M

N

O

P

Q

R

S

T

U

V

W

X

Y

Z

www.ingramcontent.com/pod-product-compliance
Lightning Source LLC
LaVergne TN
LVHW010721210926
848485LV00007B/1406/J
9781622739592